Annals of Mathematics Studies

Number 99

THE SPECTRAL THEORY OF TOEPLITZ OPERATORS

BY

L. BOUTET DE MONVEL

AND

V. GUILLEMIN

PRINCETON UNIVERSITY PRESS

AND

UNIVERSITY OF TOKYO PRESS

PRINCETON, NEW JERSEY

1981

Published in Japan exclusively by
University of Tokyo Press;
In other parts of the world by
Princeton University Press

Printed in the United States of America
by Princeton University Press, Princeton, New Jersey

Library of Congress Cataloging in Publication data will
be found on the last printed page of this book

TABLE OF CONTENTS

The Spectral Theory of Toeplitz Operators

§1. INTRODUCTION

Let W be a compact, strictly pseudoconvex domain of dimension $2n$ with smooth boundary. Let r be a smooth function on W with $r > 0$ on Int W, $r = 0$ on ∂W and $dr \neq 0$ near ∂W. Let $j : \partial W \to W$ be the inclusion map, and consider the one-form $\alpha = j^* \mathrm{Im}\, \overline{\partial r}$. It is known that α is a contact form on ∂W; that is

$$\alpha \wedge (d\alpha)^{n-1} \tag{1.1}$$

is nowhere zero on ∂W. We will denote the form (1.1) by ν. Associated with ν is a measure on ∂W which we will denote by ν as well. Let $L^2 = L^2(\partial W, \nu)$, and let H^2 be the closure in L^2 of the space of C^∞ functions on ∂W which can be extended to holomorphic functions on W. H^2 is called the space of *Hardy functions* on ∂W and the orthogonal projection $\pi : L^2 \to H^2$ is called the Szegö projector. It is not hard to show that π maps $C^\infty(\partial W)$ into $C^\infty(\partial W)$. An operator $T : C^\infty(\partial W) \to C^\infty(\partial W)$ is called a *Toeplitz operator of order* k if it can be written in the form $\pi Q \pi$ where Q is a pseudodifferential operator of order k. We will show in §2 that the set, $\mathcal{T}$, of Toeplitz operators forms a ring under composition. (This fact is not at all obvious.)

Let Σ be the subset

$$\{(x,\xi), \xi = t\alpha_x, t > 0\}$$

of $T^*\partial W$, α being the one form defined above. Σ is a *symplectic* submanifold of $T^*\partial W$: i.e. the restriction to it of the symplectic form on $T^*\partial W$ is non-degenerate. Given a Toeplitz operator, $T = \pi Q \pi$, of order k we will denote by $\sigma(T)$ the restriction of $\sigma(Q)$ to Σ where $\sigma(Q)$ is the leading symbol of Q. We will show in §2 that this definition is unambiguous, i.e. if Q_1 and Q_2 are k-th order pseudodifferential operators on ∂W and $\pi Q_1 \pi = \pi Q_2 \pi$, then $\sigma(Q_1)$ and $\sigma(Q_2)$ take the same values on Σ. We will also prove in §2 that $\sigma(T)$ satisfies the following rules:

(i) $\sigma(T_1 T_2) = \sigma(T_1)\sigma(T_2)$

(ii) $\sigma([T_1, T_2]) = \{\sigma(T_1), \sigma(T_2)\}$

(iii) If $T \in \mathcal{T}^k$ and $\sigma(T) = 0$, then $T \in \mathcal{T}^{k-1}$.

(Note that in (ii) the Poisson bracket is the intrinsic Poisson bracket on the symplectic manifold, Σ.) From (i) and (iii) we will deduce

PROPOSITION. *If* T *is a* k-th *order Toeplitz operator whose symbol is nowhere zero, there exists a* (−k)-th *order Toeplitz operator* U *such that* $TU - \pi$ *and* $UT - \pi$ *are smoothing operators.*

PROPOSITION. *If* T *is a self-adjoint Toeplitz operator of order* $k > 0$ *and* $\sigma(T)$ *is everywhere positive, then* T *has a discrete spectrum which is bounded from below and has only* $+\infty$ *as a point of accumulation.*

Let T be a Toeplitz operator of order one which is self-adjoint and has a positive symbol. Let $\lambda_1 \le \lambda_2 \le \lambda_3 \cdots$ be the spectrum of T. In this paper we will prove a number of results on the asymptotic behavior of the λ_i's. These results, which we will describe below, are analogues of results about elliptic pseudodifferential operators which are well known ([9], [12], [24]). In fact, more is true: It was shown by one of the authors (see [5]) that a pseudodifferential operator can be viewed as a special kind of Toeplitz operator; so, in fact our results *include* the results of [12], etc., as special cases.

An example of a Toeplitz operator which has no pseudodifferential counterpart is the following. Let X be a non-singular projective variety in $\mathbb{C}P^n$ which is defined, in homogeneous coordinates, by the equations $f_1(z_1, \cdots, z_{n+1}) = 0, \cdots, f_N(z_1, \cdots, z_n) = 0$. Let W be the intersection of the set of solutions of these equations with the unit disk in $\mathbb{C}^{n+1}$. Then W is a strictly pseudoconvex domain (except for a singularity at 0) which is invariant under the action of the circle group.

$$(z_1, \cdots, z_{n+1}) \to (e^{i\theta} z_1, \cdots, e^{i\theta} z_{n+1}) .$$

In particular, ∂W is a circle bundle over X. If $\partial/\partial\theta$ is the infinitesimal generator of the circle group action, the operator, $(1/\sqrt{-1})(\partial/\partial\theta)$, restricted to the Hardy space, is a self-adjoint Toeplitz operator of order one with everywhere positive symbol. Its spectrum consists of the non-negative integers, the n-th integer occurring in the spectrum with multiplicity, $p(n)$, p being the Hilbert polynomial of X. We will return to this example below.

Let T and $\{\lambda_i\}$ be as above.

THEOREM 1. *Let* $N(\lambda)$ *be the number of* λ_i*'s* $\leq \lambda$. *Then* $N(\lambda) = (\mathrm{vol}(\Sigma_1)/(2\pi)^n)\lambda^n + O(\lambda^{n-1})$, Σ_1 *being the subset of* Σ *where* $\sigma(T) \leq 1$ *and* $\mathrm{vol}(\Sigma_1)$ *its symplectic volume.*

For pseudodifferential operators, this theorem is due to Hörmander [24]. This theorem also has an amusing algebraic-geometric corollary, which is well known (see [30]):

COROLLARY. *Let* X *be a non-singular complex projective variety. Then* degree $X = \gamma_m$ volume X, γ_m *being a universal constant which depends only on the dimension,* m, *of* X.

Proof. For a projective variety the leading term in the Hilbert polynomial is $(\text{degree } X)/m!\, t^m$. Now apply the theorem above to the algebraic example described in the preceding paragraph. Q.E.D.

Let σ be the symbol of T. Since σ is a real-valued function on the symplectic manifold, Σ, there is associated with it a Hamiltonian vector field, Ξ. Since σ is homogeneous of degree one, Ξ is homogeneous of degree zero; so it actually lives on $\Sigma/\mathbf{R}^+ = \partial W$. We will think of it from now on as a contact vector field on ∂W.

THEOREM 2. *Let* Δ *be the set of cluster points of the set* $\{\lambda_i - \lambda_j\}$. *Then* $\Delta \neq \mathbf{R}$ *only if the trajectory of* Ξ *through every point of* ∂W *is periodic.*

REMARK. This result is due to Helton, [21].

This result can be somewhat sharpened: Let σ_{sub} be the subprincipal symbol of T (see §11). σ_{sub} is homogeneous of degree zero; so it can be thought of as a function on ∂W. Suppose the trajectories of Ξ are all periodic of period τ.

THEOREM 3. *Let Δ' be the set of cluster points of the set $\{e^{\sqrt{-1}\tau\lambda_i}\}$. Then the following conditions are equivalent:*

(1) *Δ' does not contain an open subset.*

(2) *Δ' is discrete.*

(3) *Δ' consists of a single point.*

(4) *The integral of σ_{sub} over any periodic trajectory, γ, of Ξ is a constant, independent of γ.*

THEOREM 4. *If one of the four equivalent conditions of Theorem 3 is satisfied there exist constants C and ρ with $C > 0$ such that the spectrum of T is entirely contained in the union of the intervals*

$$|\lambda - \tfrac{2\pi}{\tau}(m+\rho)| < C/m . \tag{1.2$_m$}$$

This theorem follows from the following more precise result:

THEOREM 5. *If σ and σ_{sub} satisfy the conditions of Theorem 3, then $T = U+V$ where U is a self-adjoint Toeplitz operator of order one whose spectrum is exactly the set $\frac{2\pi}{\tau}(m+\rho)$, $m = 0,1,2,\cdots$, V is of order -1 and U and V commute.*

REMARK. For pseudodifferential operators this result is due to Colin de Verdiére [10].

Modulo some trivial normalizations we can assume that $\tau = 2\pi$ and $\rho = 0$ in Theorem 5 (i.e. the spectrum of U is *exactly* the non-negative integers).

To describe the asymptotic behavior of the spectrum of T we are reduced, by Theorem 5, to describing the symptotic behavior of the spectrum of V on the k-th eigenspace of U. For simplicity assume that the trajectories of Ξ are all *simply* periodic of period 2π. Then we have the following Szegö-type theorem:

THEOREM 6. *Let* U *and* V *be as in Theorem 5. Let* $\{\lambda_j^{(k)}, j=1,\cdots,N_k\}$ *be the eigenvalues of* V *restricted to the* k-th *eigenspace of* U. *Let* μ_k *be the discrete measure*

$$\mu_k = 1/k^{n-1} \sum_{j=1}^{N_k} \delta(\lambda - k\lambda_j^{(k)}) .$$

Then as k *tends to* ∞, μ_k *tends to the continuous measure,* μ, *defined by*

$$\mu(f) = \gamma_n \int_{\sigma(U)=1} f(\sigma(V)(z))\,dz , \quad \textit{for} \quad f \in C_o(R) .$$

Here γ_n *is a universal constant and* dz *is the measure induced on the subset,* $\sigma(U) = 1$, *of* Σ *by the symplectic measure on* Σ.

REMARK. In the pseudodifferential setting this result is due to Weinstein, [35].

If X is a non-singular projective variety, the Toeplitz operator associated with it has spectrum concentrated on the non-negative integers (cf. supra). Applying Theorem 6 to this operator we get a rather curious Szegö-type theorem for non-singular projective varieties which, even for CP^1, seems to be new. We will discuss this theorem in §13. Concerning the multiplicity of the eigenvalues of T on the interval $(1.2)_m$ we have:

THEOREM 7. *There exists a polynomial* $p(t)$ *of degree* $n-1$ *such that the number of eigenvalues of* T *on the interval* $(1.2)_m$ *is* $p(m)$ *for* $m >> 0$.

REMARK. This theorem is also due to Colin de Verdiére, [10] in the pseudodifferential setting.

We next give a recipe for computing $p(t)$. Since the bicharacteristic flow on ∂W associated with T is simply periodic, the space of orbits is a $(2n-2)$-dimensional Hausdorff manifold, X. This manifold inherits from ∂W a canonical symplectic structure. Let α be the cohomology class of the symplectic form on X and let β be the *Todd* class of X.*

THEOREM 8. *There exists an integer* k_o *such that the polynomial,* p, *in Theorem 7 is given by the formula*

$$p(t+k_o) = e^{-t\alpha}\beta[X] . \tag{1.3}$$

REMARK. If T is the Toeplitz operator associated with an algebraic variety, as described above, p is just the Hilbert polynomial of the variety, and (1.3) can be found in Hirzebruch, [22].

The preceding theorems (Theorems 4-8) apply to a rather special class of Toeplitz operators. It is clear that, generically, the bicharacteristic vector field, Ξ, has relatively few periodic trajectories. Theorem 2 shows that even if *one* trajectory is non-periodic then the spectrum of T is rather equidistributed and, in a certain sense, rather devoid of structure. It turns out, however, that if Ξ is sufficiently generic, geometric data about T can be extracted from the spectrum by means of the generating function:

$$e(t) = \text{trace } e^{\sqrt{-1}tT} = \sum e^{\sqrt{-1}\lambda_i t} . \tag{1.4}$$

THEOREM 9. *The function* $e(t)$ *is a tempered distribution. If* τ *is in the singular support of* e *there exists a periodic trajectory,* γ, *of* Ξ *of period* $\tau_\gamma = \tau$.

---

*Associated with the symplectic structure on X is an equivalence class of almost complex structures (see [33]). The Todd class of X is the Todd class of any representative almost-complex structure in this equivalence class.

THEOREM 10. *Suppose all periodic trajectories,* γ, *of* Ξ *are non-degenerate in the sense that the Poincaré map,* P_γ, *attached to* γ *has no eigenvalues equal to one. Then*

$$e(t) = \sum_{\gamma} \frac{\tau_\gamma^\#}{|I - P_\gamma|^{1/2}} \left(e^{i\int_\gamma \sigma_{sub}}\right)(t - \tau_\gamma + io+)^{-1} + e_1(t) \tag{1.5}$$

where $e_1(t) \in L^1_{loc}(R^+)$. *Here* τ_γ *is the period of* γ, $\tau_\gamma^\#$ *its primitive period and* $\int_\gamma \sigma_{sub}$ *the integral of the subprincipal symbol of* P *over* γ.

We will now give a brief outline of how these theorems are proved. In some sense the key result above is the trace formula (1.5). It turns out that most of the other results can be derived from a generalized form of (1.5) by Tauberian arguments just as in §1 of [12]. The formula (1.5) and its generalizations are proved by making use of functorial properties of distributional kernels like those used to define Szegö projectors. In dealing with such distributions we were faced with two options: either to treat them as Fourier integral distributions of complex type in the sense of Melin-Sjöstrand [28] or as Hermite distributions in the sense of Boutet de Monvel [4]. We finally settled on the second option. Its advantages are that the symbol calculus for Hermite distributions reflects only what goes on the real wave front sets of the distributions involved, not on their "almost-complexifications." For instance the clean intersection hypotheses in Theorem 12.9 would, in the complex Fourier integral distribution approach, have to be strengthened to include assumptions about the intersections at complex points, assumptions which seem unduly restrictive. (We will, however, use the Fourier integral operator calculus of Melin and Sjöstrand in the appendix.)

As we mentioned above, in §2 we will discuss elementary properties of Toeplitz operators, proving for instance that they form a ring under composition. In §3 we will review general facts about oscillatory integrals of Hermite type. Much of this material is taken from Guillemin, [15]. In §§4-7 we discuss the symbols of these distributions and in §8 prove our basic composition formula. In §§9-11 we discuss applications of this formula (in particular, applications to Toeplitz operators), and, in §12, we derive from our composition formula the trace formula alluded to above. In §13 we use the trace formula to read off several of the spectral theorems listed above and indicate sketchily how the rest are proved. §14 is devoted to the proof of Theorem 8. The proof of this theorem requires looking at a class of operators which are not, strictly speaking, Toeplitz though they resemble Toeplitz operators micro-locally. For this reason, as well as for the reason that these operators are interesting in their own right, all the results of this paper will be formulated so as to apply to these "generalized" Toeplitz operators.

In §15 we will discuss some connections between our results and some recent work of Menikoff and Sjöstrand on the spectral properties of hypoelliptic operators with quadratically degenerating symbols. Finally we have inserted at the end a longish appendix whose purpose is two-fold: a) to show that if X is a compact manifold and Σ a homogeneous symplectic submanifold of T^*X-0, then Σ possesses a "generalized Toeplitz structure" in the sense of §2; and b) to show that such a Toeplitz structure can be described by a micro-local analogue of the $\bar{\partial}_b$-complex. Both a) and b) are required for the proof of Theorem 8 and are also interesting in their own right.

We would like to express our warmest thanks to Lars Hörmander, Richard Melrose, Arthur Menikoff, Edward Miller and Johannes Sjöstrand for their advice and encouragement. We would also like to thank the mathematics departments of the Institute for Advanced Study and Princeton University for their sponsorship of this work. This monograph was written in the Spring of 1978 while one of the authors was visiting at Princeton and the other at I.A.S.

§2. GENERALIZED TOEPLITZ OPERATORS

In [8] Boutet de Monvel and Sjöstrand exhibit an operator $\pi : L^2(\mathbf{R}^n) \to L^2(\mathbf{R}^n)$ which has all the essential micro-local features of the Szegö projector. To define this operator let $\mathbf{R}^n = \mathbf{R}^p \times \mathbf{R}^q$ with $x = (y, t)$ as coordinates and $\xi = (\eta, \tau)$ as dual coordinates. Consider the system of operators $D_j : \mathcal{S}(\mathbf{R}^n) \to \mathcal{S}(\mathbf{R}^n)$, $j = 1, \cdots, p$ defined by

$$D_j = (1/\sqrt{-1})(\partial/\partial y_j + y_j |D_t|) \tag{2.1}$$

where $|D_t|$ is, as usual, the operator $\widehat{|D_t| f} = |\tau| \hat{f}$. Let H be the L^2-closure of the space of solutions: $\{f \in \mathcal{S}, D_j f = 0, j = 1, \cdots, p\}$, and let π be the orthogonal projector of L^2 onto H.

PROPOSITION 2.1. *The following oscillatory integral is the Schwartz kernel of* π:

$$(2\pi)^{-q} \int e^{i(t-t')\tau - (|y|^2 + |y'|^2)|\tau|} (|\tau|/\pi)^{p/2} d\tau \,. \tag{2.2}$$

Proof. Performing a partial Fourier transform in the t-variable on $L^2(\mathbf{R}^n)$ π gets conjugated into the operator

$$(\pi' g)(y, \tau) = \int e^{-\frac{1}{2}(|y|^2 + |y'|^2)|\tau|} (|\tau|/\pi)^{p/2} g(y', \tau)\, dy' \tag{2.3}$$

and the operator (2.1) gets conjugated into the operator

$$D'_j = (1/\sqrt{-1})(\partial/\partial y_j + y_j |\tau|) \qquad j = 1, \cdots, p \,. \tag{2.4}$$

The system (2.4) has for fixed τ the solution

$$e^{-\frac{1}{2}|y|^2|\tau|}(|\tau|/\pi)^{p/4}\ . \tag{2.5}$$

For fixed τ (2.5) has L^2-norm one as a function of y and is, up to unit multiples, the only solution of (2.4) with this property. Therefore, for fixed τ, the operator with kernel

$$e^{-\frac{1}{2}(|y|^2+|y'|^2)|\tau|}(|\tau|/\pi)^{p/2} \tag{2.6}$$

is orthogonal projection onto the kernel of (2.4). Now compare (2.6) and (2.3). Q.E.D.

Let $R: L^2(\mathbf{R}^q) \to L^2(\mathbf{R}^n)$ be the operator

$$Rf(y,t) = (2\pi)^{-q} \int e^{it\tau - \frac{1}{2}|y|^2|\tau|}(|\tau|/\pi)^{p/4} f(\tau)\, d\tau\ . \tag{2.7}$$

PROPOSITION 2.2. R *maps* $L^2(\mathbf{R}^q)$ *isomorphically onto* H.

Proof. Conjugating with respect to Fourier transform in the t-variable R becomes the operator

$$(R'f)(y,\tau) = e^{-\frac{1}{2}|y|^2|\tau|}(|\tau|/\pi)^{p/4} f(\tau)$$

which maps $L^2(\mathbf{R}^q)$ isometrically onto the space of solutions of (2.4). Q.E.D.

As a corollary we get

$$R^*R = I \quad \textit{and} \quad RR^* = \pi\ . \tag{2.8}$$

DEFINITION 2.3. A *Toeplitz* operator on $\mathbf{R}^p \times \mathbf{R}^q$ is an operator of the form $T = RQR^*$ where $Q: L^2(\mathbf{R}^q) \to L^2(\mathbf{R}^q)$ is a pseudodifferential operator.

The set of Toeplitz operators forms a ring which is isomorphic to the ring of pseudodifferential operators on $L^2(\mathbf{R}^q)$.

DEFINITION 2.4. Let $T = RQR^*$ be a Toeplitz operator. We will say that T is *of order* k if Q is of order k.* The *symbol* of T, $\sigma(T)$, is $\sigma(Q)$, viewed as a function on the subset $y = 0$, $\eta = 0$ of $T^*\mathbf{R}^n$. Similarly, $\sigma_{sub}(T)$, the *subprincipal symbol* of T is just $\sigma_{sub}(Q)$. The following formulas are immediate:

$$(2.9)_a \qquad \sigma(T_1 T_2) = \sigma(T_1)\sigma(T_2)$$

$$(2.9)_b \qquad \sigma([T_1, T_2)] = [\sigma(T_1), \sigma(T_2)]$$

$$(2.9)_c \qquad \sigma_{sub}(T_1 T_2) = \sigma(T_1)\sigma_{sub}(T_2) + \sigma(T_2)\sigma_{sub}(T_1) + (1/2\sqrt{-1})\{\sigma(T_1), \sigma(T_2)\}\ .$$

PROPOSITION 2.3. *Let* $Q : \mathcal{S}(\mathbf{R}^n) \to L^2(\mathbf{R}^n)$ *be a pseudodifferential operator of order* k. *Then* $R^*QR : L^2(\mathbf{R}^q) \to L^2(\mathbf{R}^q)$ *is a pseudodifferential operator of order* k, *and its symbol is just the restriction of the symbol of* Q *to the set* $y = \eta = 0$.

Proof. This is an immediate corollary of some general theorems about composition of F.I.O.'s of Hermite type which we will discuss in §9.

COROLLARY. *Let* $Q : L^2(\mathbf{R}^n) \to L^2(\mathbf{R}^n)$ *be a pseudodifferential operator. Then* $\pi Q \pi$ *is a Toeplitz operator, whose symbol is equal to the symbol of* Q *restricted to the set* $y = \eta = 0$.

PROPOSITION 2.4. *Let* $\mathcal{R}(x, t)$ *be the Schwartz kernel of the operator* R. *Then the wave front set of* $\mathcal{R}$ *consists of the points* $\{(x, \xi, t, -\tau)\}$ *where* (x, ξ, t, τ) *lies on the graph of the symplectic imbedding* $\phi : T^*\mathbf{R}^q \to T^*\mathbf{R}^n$, $\phi(t, \tau) = (0, 0, t, \tau)$.

Proof. $\mathcal{R}$ is the oscillatory integral:

$$\mathcal{R}(y, t, t') = \int e^{i(t-t')\tau - \frac{1}{2}|y|^2|\tau|}(|\tau|/\pi)^{p/4} d\tau$$

---

*Strictly speaking we should assume $k \leq 0$ and require that Q map L^2 into L^2; however we will deliberately be a little vague on this point.

which is smooth except when $y = 0$ and $t = t'$. For t' fixed, $\mathfrak{R}$ satisfies the equations, $D_j \mathfrak{R} = 0$, and it also satisfies the equations $(1/\sqrt{-1})(\partial/\partial t_i - \partial/\partial t'_i)\mathfrak{R} = 0$ with $i = 1, \cdots, q$; so the wave front set is contained in the set where the symbols of these equations vanish. Q.E.D.

REMARK. Proposition 2.4 also follows from properties of oscillatory integrals with quadratically damped amplitudes which we will describe in §3.

COROLLARY. *Let* Σ *be the subset,* $y = \eta = 0$, *of* T^*R^n (*i.e.* Σ *is the image of the imbedding* ϕ). *Then the wave front set of* $S.K.\pi$ *is the set* $\{(x,\xi,x,-\xi),(x,\xi) \in \Sigma\}$.

Proof. This follows from the formula (2.8). It also follows from general facts about oscillatory integrals such as (2.2). Q.E.D.

The following is in some sense the main result of this section:

PROPOSITION 2.5. *Let* $T: H \to H$ *be a Toeplitz operator. Then there exists a pseudodifferential operator* $Q: L^2(R^n) \to L^2(R^n)$ *such that* Q *maps* H *into* H *and the restriction of* Q *to* H *is* T.

The proof requires some preliminary lemmas.

LEMMA 2.6. *Let* $Q: L^2(R^n) \to L^2(R^n)$ *be a pseudodifferential operator whose total symbol vanishes to infinite order on* Σ. *Then there exist pseudodifferential operators* $L_1, \cdots, L_p$ *with the same property such that* $Q - \Sigma L_j D_j$ *is smoothing.*

Proof. Let $\square = \Sigma D_j^* D_j$. The symbol of $\square$ at (x,ξ) is just the square of the distance from (x,ξ) to Σ. Since $\sigma(Q)$ vanishes to infinite order on Σ, $\sigma(Q)/\sigma(\square)$ is a smooth function which also vanishes to infinite order on Σ. Let P_o be a pseudodifferential operator having this function as its leading symbol. We can arrange that the total symbol of P_o vanishes to infinite order on Σ. Then $Q_1 = Q - P_o\square$ is of order one less than Q and its total symbol vanishes to infinite order on Σ. Repeating the above argument with Q_1, etc., we can recursively construct an operator P such that the total symbol of P vanishes to infinite order on Σ and $Q - P\Box$ is smoothing. Now set $L_j = PD_j^*$. Q.E.D.

LEMMA 2.7. *There exist operators* $M_j : (\mathcal{S}(R^n) \to L^2(R^n)$ *such that* $\mathrm{Id} = \pi + \Sigma M_j D_j$.

Proof. See [8].

LEMMA 2.8. *Given a Toeplitz operator,* T, *there exists a pseudodifferential operator* $Q : L^2(R^n) \to L^2(R^n)$ *such that* $\sigma(Q)$, *restricted to* Σ, *is equal to* $\sigma(T)$ *and, for all* $j = 1, \cdots, p$, *the total symbol of* $[D_j, Q]$ *vanishes to infinite order on* Σ.

Proof. If q_o is the leading symbol of Q then the leading symbol of $[D_j, Q]$ is $Z_j q_o$ where Z_j is the vector field

$$Z_j = (1/\sqrt{-1})(\partial/\partial y_j) + (|\tau|(\partial/\partial \eta_j) - y_j \sum_{\alpha=1}^{q} (\tau_\alpha/|\tau|)(\partial/\partial t_\alpha)) . \tag{2.10}$$

We first arrange that $\sigma(Q)|\Sigma = \sigma(T)$ and $\sigma([D_j, Q])$ vanishes to infinite order on Σ. This amounts to solving the overdetermined boundary value problem

$$Z_j q_o = 0, \quad j = 1, \cdots, p, \quad q_o(0, 0, t, \tau) = \sigma(T)(t, \tau),$$

formally around $y = \eta = 0$. Since the Z_j's commute and each Z_j has the form (2.10) this is clearly possible. Consider now the subprincipal symbol of $[D_j, Q]$ near Σ. Let this subprincipal symbol be r_j. By Jacobi's identity

$$[D_j, [D_k, Q]] = [D_k, [D_j, Q]]$$

since $[D_j, D_k] = 0$. From this identity we conclude that $Z_j r_k - Z_k r_j$ vanishes to infinite order on Σ. Therefore by the Poincaré lemma we can find q_1 such that $Z_j q_1 - r_j$ vanishes to infinite order on Σ for $j = 1, \cdots, p$. If we modify Q by adding to it an operator with symbol q_1, we now are in a situation where both the symbol and the subprincipal symbol of $[D_j, Q]$ vanish to infinite order on Σ. It is clear how to continue this process. Q.E.D.

We will now prove Proposition 2.5. Given T we first choose Q so that it satisfies the hypotheses of Lemma 2.8. By Lemma 2.6 there exist L_j^k such that

$$D_jQ - QD_j - \sum L_j^k D_k$$

is smoothing. Thus $D_jQ\pi$ is smoothing. By Lemma 2.7, $Q\pi - \pi Q\pi$ is smoothing. By hypothesis the symbol of $\pi Q\pi$ is the same as the symbol of T.

Replacing T by $T - \pi Q\pi$, etc., one can construct recursively a Q_1 such that both $\pi Q_1\pi - T$ and $Q_1\pi - \pi Q_1\pi$ are smoothing. Let $S = \pi Q_1\pi - Q_1\pi$ and let $Q_2 = Q_1 + S$. Then $Q_2\pi = Q_1\pi + S\pi = Q_1\pi + S = \pi Q_1\pi = \pi(Q_1 + S)\pi = \pi Q_2\pi$ since $\pi S\pi = 0$. Thus Q_2 preserves H. Finally $W = T - Q_2\pi$ is smoothing. Let $Q_3 = Q_2 + W$. Then Q_3 preserves H and its restriction to H is T. Q.E.D.

Our next result is a sharpening of Proposition 2.5.

THEOREM 2.9. *Let* T *be a self-adjoint Toeplitz operator. Then there exists a self-adjoint pseudodifferential operator* $Q : L^2(\mathbf{R}^n) \to L^2(\mathbf{R}^n)$ *such that* Q *preserves* H *and the restriction of* Q *to* H *is* T.

Proof. Let Q_o be any pseudodifferential operator satisfying the hypotheses of Proposition 2.5 and let $p_o(x,\xi) = \operatorname{Im} \sigma(Q_o)(x,\xi)$. Since $\sigma(Q_o) = \sigma(T)$ on Σ and $\sigma(T)$ is real valued, $p(x,\xi)$ vanishes on Σ. Thus there exist smooth real valued functions f_j and g_j, $j = 1, \cdots, p$, such that for $|\tau| = 1$

$$p_o(x,\xi) = \sum - f_j y_j + g_j \eta_j = \operatorname{Im} \sum (f_j + \sqrt{-1}\, g_j)(\eta_j - \sqrt{-1}\, y_j)\,.$$

Extending f_j and g_j to the set $\tau \neq 0$ so that they are homogeneous of one less degree than p_o we have

$$p_o(x,\xi) = \sum (f_j + \sqrt{-1}g_j)(\eta_j - |\tau|\sqrt{-1}y_j) = \operatorname{Im}\sum h_j\sigma(D_j)$$

where $h_j = f_j + \sqrt{-1}g_j$. Now let H_j, $1,\cdots,p$, be pseudodifferential operators with $\sigma(H_j) = h_j$ in a neighborhood of Σ, and consider $Q_1 = Q_o - \Sigma H_j D_j$. This operator preserves H and the symbol of $Q_1 - Q_1^*$ is of order one less than the symbol of Q on a neighborhood of Σ. Repeating the argument above with p_o replaced by $p_1 = (1/2\sqrt{-1})\sigma(Q_1 - Q_1^*)$ we can find pseudodifferential operators H_j', $j = 1,\cdots,p$, of order two less than the order of Q such that if $Q_2 = Q_1 - \Sigma H_j D_j$ then $Q_2 - Q_2^*$ is of order two less than Q near Σ. It is clear that by repeating this procedure infinitely often we can manufacture an operator, Q_∞, such that Q_∞ preserves H, $\sigma(Q_\infty) = \sigma(Q_o)$ on Σ and $Q_\infty - Q_\infty^*$ is smoothing in a neighborhood of Σ. Let $Q' = (Q_\infty + Q_\infty^*)/2$. Q' is self-adjoint; so $T_1 = T - \pi Q'\pi$ is self-adjoint and of order one less than Q. Repeating the above argument with T_1, etc., one can manufacture recursively a $\tilde{Q}$ such that $\tilde{Q} - \tilde{Q}^*$ is smoothing, $\tilde{Q}\pi - \pi\tilde{Q}\pi$ is smoothing and $T - \pi\tilde{Q}\pi$ is smoothing. Replacing $\tilde{Q}$ by $(\tilde{Q} + \tilde{Q}^*)/2$ we can arrange that $\tilde{Q}$ is self-adjoint, and replacing $\tilde{Q}$ by $\pi\tilde{Q}\pi + (1-\pi)\tilde{Q}(1-\pi)$ we can arrange that it is self-adjoint and preserves H. Finally set $Q = \tilde{Q} + T - \pi\tilde{Q}\pi$. Q.E.D.

COROLLARY. *If* T *is a Toeplitz operator there exists a pseudodifferential operator on* R^n, Q, *such that* $[Q,\pi] = 0$ *and* $T = Q$ *restricted to* H.

Proof. Let $T_1 = (T + T^*)/2$ and $T_2 = (T - T^*)/2\sqrt{-1}$. T_1 and T_2 are self-adjoint and $T = T_1 + \sqrt{-1}\,T_2$. If Q_1 and Q_2 are self-adjoint extensions of T_1 and T_2 then $Q = Q_1 + \sqrt{-1}\,Q_2$ has the required properties. Q.E.D.

We now describe how these results can be extended from R^n to manifolds. Let X be a compact C^∞ manifold and let $L^2(X)$ be the Hilbert space of square integrable half-densities on X. Let Σ be a closed conic submanifold of $T^*X - 0$ with the property that the symplectic form, restricted to it, is non-degenerate.

DEFINITION 2.10. A Toeplitz structure on Σ is an operator $\pi_\Sigma : L^2(X) \to L^2(X)$ such that

I. $\pi_\Sigma^2 = \pi_\Sigma$ and $\pi_\Sigma^* = \pi_\Sigma$.

II. $WF(\pi) = \{(x,\xi,x,-\xi),(x,\xi) \in \Sigma\}$.

III. For each $(x,\xi) \in \Sigma$ there exists a conic neighborhood U of (x,ξ) in $T^*X - 0$, a conic open set V in $T^*R^n - 0$, a homogeneous canonical transformation $\phi : U \to V$ and a Fourier integral operator $F : L^2(X) \to L^2(R^n)$ extending ϕ, such that $F^*F - I$ is C^∞ on U and $F\pi_\Sigma F^* - \pi$ is C^∞ on V.

It was proved by Boutet de Monvel and Sjöstrand in [7] that the Szego projector, discussed in §1, satisfies these axioms. It will be proved in §4 of the appendix that for every compact manifold X and every symplectic cone, Σ, in $T^*X - 0$ there exists a Toeplitz structure on Σ. This Toeplitz structure is not unique, but is in a certain sense unique up to homotopy. We will have more to say about this result in §14.

Let H_Σ be the image of π_Σ in $L^2(X)$. It follows from condition II that π_Σ maps $C^\infty(X)$ into $C^\infty(X) \cap H_\Sigma$. By a *Toeplitz operator on* H_Σ. we will mean an operator $T : C^\infty(X) \cap H_\Sigma \to C^\infty(X) \cap H_\Sigma$ of the form $\pi_\Sigma Q \pi_\Sigma$ where $Q : C^\infty(X) \to C^\infty(X)$ is a pseudodifferential operator. We will say that T is of order k if Q is of order k.

PROPOSITION 2.11. *Suppose* $T = \pi_\Sigma Q_i \pi_\Sigma$, $i = 1,2$ *where* Q_1 *and* Q_2 *are* k-th *order pseudodifferential operators. Then* $\sigma(Q_1)|\Sigma = \sigma(Q_2)|\Sigma$.

Proof. It is enough to prove this for the canonical model. For the canonical model, however, this follows from Proposition 2.3. Q.E.D.

DEFINITION 2.11. If T is a Toeplitz operator of order k on Σ and $T = \pi_\Sigma Q \pi_\Sigma$ with Q of order k, then $\sigma(T) = \sigma(Q)|\Sigma$.

PROPOSITION 2.12. *If* T *is of order* k *and* $\sigma(T) = 0$ *then* T *is of order* $k-1$.

Proof. Again it is enough to verify this for the canonical model. For the canonical model, however, this follows from Proposition 2.5 and Proposition 2.3. Q.E.D.

PROPOSITION 2.13. *Let* T *be a Toeplitz operator on* Σ *of order* k. *Then there exists a pseudodifferential operator,* Q, *of order* k *such that* $T = \pi_\Sigma Q \pi_\Sigma$ *and* $[\pi_\Sigma, Q] = 0$.

Proof. By a pseudodifferential partition of unity we can assume that $T = \pi_\Sigma Q_o \pi_\Sigma$ where Q_o is C^∞ except in an arbitrarily small conic neighborhood. Conjugating π_Σ with π and making use of Theorem 2.9 we can easily find a Q_1 such that $T - \pi_\Sigma Q_1 \pi_\Sigma$ is smoothing and $[\pi_\Sigma, Q_1]$ is smoothing. Now let $Q_2 = \pi_\Sigma Q_1 \pi_\Sigma + (1-\pi_\Sigma) Q_1 (1-\pi_\Sigma)$. Q_2 differs from Q_1 by a smoothing operator and $[\pi_\Sigma, Q_2] = 0$. Finally set $Q = Q_2 + T - \pi_\Sigma Q_1 \pi_\Sigma$. Q.E.D.

COROLLARY 1. *Toeplitz operators form a ring under composition, and if* T *is a Toeplitz operator its transpose,* T^*, *is a Toeplitz operator.*

COROLLARY 2. *The symbol,* σ, *defined above satisfies the following composition rules.*

$$\sigma(T_1 T_2) = \sigma(T_1)\sigma(T_2) \tag{2.11$_a$}$$

$$\sigma([T_1, T_2]) = \{\sigma(T_1), \sigma(T_2)\}_\Sigma . \tag{2.11$_b$}$$

COROLLARY 3. *Let* T *be a Toeplitz operator of order* k. *Suppose* $\sigma(T)$ *is nowhere zero on* Σ. *Then there exists a Toeplitz operator* U *of order* $-k$ *such that* $TU - \pi_\Sigma$ *and* $UT - \pi_\Sigma$ *are smoothing.*

Proof. Let Q be a pseudodifferential operator such that $T = \pi_\Sigma Q \pi_\Sigma$ and $[\pi_\Sigma, Q] = 0$. Since $\sigma(Q) \neq 0$ on Σ, there exists a pseudodifferential operator, R, of order $-k$ such that $QR - I$ and $RQ - I$ are smoothing in a neighborhood of Σ. Set $U = \pi_\Sigma R \pi_\Sigma$. Q.E.D.

PROPOSITION 2.14. *Let* T *be a self-adjoint Toeplitz operator of order* >0. *Suppose* $\sigma(T)>0$ *everywhere on* Σ. *Then the spectrum of* T *is discrete, bounded from below and has only* $+\infty$ *as a point of accumulation.*

Proof. Let Q be a pseudodifferential operator such that $T=\pi_\Sigma Q \pi_\Sigma$ and $[Q,\pi_\Sigma]$ is smoothing. Since $\operatorname{Re}\sigma(Q)>0$ near Σ we can assume $\operatorname{Re}\sigma(Q)>0$ everywhere. Replacing Q by $(Q+Q^*)/2$ we can assume Q is self-adjoint and $\sigma(Q)$ everywhere positive. Finally Q differs from $\pi_\Sigma Q \pi_\Sigma + (1-\pi_\Sigma)Q(1-\pi_\Sigma)$ by a smoothing operator; so we can assume Q has all of the above properties and also commutes with π_Σ. Now Q is a self-adjoint elliptic pseudodifferential operator of positive degree with a symbol which is everywhere positive; so its spectrum is discrete, bounded from below and has only $+\infty$ as a point of accumulation. The same is therefore true of the restriction of Q to the invariant subspace, H_Σ.

Q.E.D.

§3. FOURIER INTEGRAL OPERATORS OF HERMITE TYPE

Given a smooth manifold, X, and a homogeneous isotropic submanifold, Σ, of the cotangent bundle of X one can define a class of distributions with wave front set concentrated on Σ having many of the properties of the Fourier integral distributions of Hörmander [25]. Indeed when Σ is Lagrangian (i.e. maximal isotropic) these distributions are just the usual Fourier integral distributions. These distributions, which we will define below, are called Fourier integral distributions of *Hermite type*. They were first studied systematically in [4] and their symbolic properties explored in [15]. The next few sections of this paper will be partly a review of the material in [4] and [15] and partly an elaboration of the symbolic theory of [15] with a view to the applications in §13.

Let $\mathbf{R}^N = \mathbf{R}^k \times \mathbf{R}^\ell$ with coordinates $\theta = (\tau, \eta)$, and let X be an open subset of $\mathbf{R}^n$. Let $a(x, \tau, \eta)$ be a smooth function on $X \times \mathbf{R}^N$. We will say that a belongs to the symbol class $\mathcal{H}^m(k, \ell)$ if for all numbers $r > 0$ all multi-indices α, β and γ and all $K \subset\subset X$ there exists a constant $C > 0$ such that

$$(3.1) \qquad |D_x^\alpha D_\tau^\beta D_\eta^\gamma a(x, \tau, \eta)| \leq C |\tau|^{m-|\gamma|} (1 + |\eta|)^{-r}$$

for all $x \epsilon K$. Notice that if a is in $\mathcal{H}^m(k, \ell)$ then as a function of η it is rapidly decreasing and as a function of τ it behaves as a classical symbol of type $S^m_{1,0}$. Let $a_i(x, \tau, \eta)$, $i = 0, 1, 2, \cdots$, be a sequence of elements of $\mathcal{H}^{m_i}(k, \ell)$ with $m_i \to -\infty$, $a_i(x, \tau, \eta)$ being homogeneous of degree m_i in τ for $|\tau| >> 0$. Given an element, a, of $\mathcal{H}^{m_o}(k, \ell)$ we will say that

$$(3.2) \qquad a(x, \tau, \eta) \sim \sum a_i(x, \tau, \eta)$$

if $a(x,\tau,\eta) - \sum_{j<i} a_j(x,\tau,\eta) \in \mathcal{H}^{m_i}(k,\ell)$ for all i. Unless otherwise stated we will usually assume below that the m_i's in (3.2) are *either integers or half-integers.*

Let $\phi : X \times \mathbf{R}^N - 0 \to \mathbf{R}$ be a non-degenerate phase function in the sense of Hörmander, [25]. Assume the critical set of ϕ,

$$C_\phi = \{(x,\theta), \partial\phi/\partial\theta_1 = \cdots = \partial\phi/\partial\theta_N = 0\}$$

intersects the set, $\eta_1 = \cdots = \eta_\ell = 0$, transversally. Let Σ be the image in $T^*X - 0$ of this intersection under the mapping

$$(x,\theta) \to (x, \partial\phi/\partial x), \quad C_\phi \to T^*X - 0 . \tag{3.3}$$

Σ is a homogeneous, isotropic submanifold of $T^*X - 0$ of dimension $n-\ell$.

Consider now the oscillatory integral

$$I_a = \int a(x,\tau,\eta/\sqrt{|\tau|})e^{i\phi(x,\theta)}d\theta \tag{3.4}$$

where $a \in \mathcal{H}^{m-N/2}(k,\ell)$. We will prove below.

PROPOSITION 3.1. *The wave front set of (3.4) is contained in* Σ.

We will say that the pair $(X \times \mathbf{R}^N, \phi)$ is a *parametrization* of Σ if Σ is the image of $\eta_1 = \cdots = \eta_\ell = 0$ under the map (3.3). We will also prove below.

PROPOSITION 3.2. *If* $(X \times \mathbf{R}^N, \phi)$ *and* $(X \times \mathbf{R}^{N'}, \phi')$ *are two different parametrizations of* Σ *then for every* $p \in \Sigma$ *an oscillatory integral of the form (3.4) can also be written as an oscillatory integral of the form*

$$\int a'(x,\tau',\eta'/\sqrt{|\tau'|})e^{i\phi'(x,\tau',\eta')}d\tau' d\eta' + h .$$

Here $a' \in \mathcal{H}^{m'}(k',\ell)$, $m' + N'/2 = m + N/2$ *and* $p \notin WF(h)$.

To prove these two propositions we will first give an alternative description of the space of distributions defined by (3.4). Let $X = \mathbf{R}^n = \mathbf{R}^k \times \mathbf{R}^\ell$ with coordinates $x = (t,y)$ and let $\xi = (\tau,\eta)$ be the dual coordinates. Let Σ_o be the subset $x = y = \eta = 0$ of $T^*\mathbf{R}^n - 0$ and let ϕ be the phase function $\phi(x,t,\tau,\eta) = \tau t + \eta y$ on $\mathbf{R}^n \times \mathbf{R}^n$. ϕ is clearly parametrizing phase function for Σ_o. In [16] the following analogues of the propositions above were proved.

PROPOSITION 3.1′. *Let* I_b *be the oscillatory integral*

$$\int b(\tau, \eta/\sqrt{|\tau|}) e^{i(t\tau + y\eta)} d\tau\, d\eta \qquad \text{with} \qquad b \in \mathcal{H}^m(k,\ell)\,. \tag{3.5}$$

Then $WF(I_b) \subset \Sigma_o$.

PROPOSITION 3.2′. *Let* $f: T^*\mathbf{R}^n - 0 \to T^*\mathbf{R}^n - 0$ *be a homogeneous canonical transformation mapping* Σ_o *into* Σ_o *and let* F *be a Fourier integral operator of order* r *whose underlying canonical transformation is* f. *Then there exists a symbol* $b'(\tau,\eta) \in \mathcal{H}^{m+r}(k,\ell)$ *such that*

$$FI_b = I_{b'} + h \tag{3.6}$$

with $h \in C^\infty$.

Now let X be an n-dimensional manifold and Σ a homogeneous isotropic submanifold of $T^*X - 0$ of dimensional $n - \ell$. Given $p \in \Sigma$ and the point $p_o : x = 0$, $t = 0$, $\tau = (1,0,\cdots,0)$, $\eta = 0$, in Σ_o there exists a homogeneous canonical transformation, f, mapping a neighborhood of p_o onto a neighborhood of p and mapping (Σ_o, p_o) onto (Σ, p). (See §3 of [15].) Let F be a zeroth order elliptic Fourier integral operator with f as its underlying canonical transformation and let I_b be a distribution of type (3.5) with the support of b contained in a small conic neighborhood of $\eta = 0$, $\tau = (1,0,\cdots,0)$. By Proposition 3.1′, FI_b has its wave-front set concentrated in a small conic neighborhood of p in Σ. Moreover, it is clear from (3.6) that if f' and F' have the same properties as f and F then there exists a $b'(\tau,\eta) \in \mathcal{H}^m(k,\ell)$ such that $FI_b - F'I_{b'}$ is smooth near p. Therefore the space $\{FI_b, b(\tau,\eta) \in \mathcal{H}^m(k,\ell)\}$ is intrinsically defined modulo C^∞. We will now show that FI_b is an oscillatory integral of type 3.3. Let $\phi(x,\xi)$ be a generating function for the canonical transformation, f. This means that ϕ is homogeneous of degree one in ξ, the matrix $(\partial\phi/\partial x_i\partial\xi_j)$ is non-singular and f is obtained from ϕ by solving the system of equations

$$\begin{aligned} \xi'_i &= (\partial\phi/\partial x_i)(x',\xi) \\ x_i &= (\partial\phi/\partial\xi_i)(x',\xi) \end{aligned} \tag{3.7}$$

for (x',ξ') in terms of (x,ξ) near $(x_o,\xi_o) = p_o$. Let $\rho(x,\tau)$ be smooth, homogeneous of degree zero in τ for $|\tau| >> 0$, equal to one near p_o and supported in a small conic neighborhood of p_o in Σ_o. Then by Egorov, [14], the operator $F: C_o^\infty(R^n) \to C^\infty(X)$ defined by

$$Fu(x) = \int \rho(x,\tau)e^{i\phi(x,\xi)}\hat{u}(\xi)d\xi$$

is a Fourier integral operator with f as its underlying canonical transformation. With this choice of F we have

$$FI_b = \int \rho(x,\tau)e^{i\phi(x,\xi)}b(\tau,\eta/\sqrt{|\tau|})d\tau\, d\eta\ . \tag{3.8}$$

By (3.7) the image of Σ_o is just the set

$$\eta_1 = \cdots = \eta_\ell = 0\,, \quad \partial\phi/\partial\xi_1 = \cdots = \partial\phi/\partial\xi_n = 0\ ;$$

so ϕ is a parametrizing phase function for Σ. This shows that (3.8) is an oscillatory integral of the form (3.4). Conversely if ϕ is the phase function (3.7) and I_a is an oscillatory integral of the form (3.4) then $I_a = FI_b + h$ with $p \notin WF(h)$ and $b(\tau,\eta) \in \mathcal{H}^m(k,\ell)$. (See [15].)

It is clear that the phase function, ϕ, is already fairly close to being the most general type of phase function possible. In fact any phase function $\phi(x, \xi)$ having the following two properties can arise as the generating function of a canonical transformation:

(3.9) a) The number of phase variables $\xi_1, \cdots, \xi_n$ is equal to the number of base variables, $x_1, \cdots, x_n$.

(3.9) b) The critical set: $\partial\phi/\partial\xi = 0$ is transversal to the set $\xi_1 = \cdots = \xi_n = 0$.

Let us show that given an arbitrary phase function ϕ we can modify it so that if ϕ' is the modified phase function, then ϕ and ϕ' define the same class of oscillatory integrals and ϕ' satisfies (3.9). First of all if ϕ has less than n phase variables we can modify it by adding on the term $(\tau_{k+1}^2 + \cdots + \tau_n^2)/|\theta|$ as in §3 of Hörmander [25]. Just as in §3 of [25] one can see that this will not change the form of the amplitude in (3.4). If ϕ has more than n phase variables, the superfluous phase variables must be among the τ's because of the transversality condition on the η's; so we can reduce the number of phase variables by applying stationary phase to certain of the τ's. Again as in §3 of [25] it is easy to see that this does not affect the form of the amplitude in (3.4). Finally if ϕ fails to satisfy $(3.9)_b$ then because the η's are already transversal to C_ϕ we can make a linear change of variables of the form $\tau_i' = \tau_i - |\tau| \sum a_{ij} x_j$, so that (τ', η) is transversal to the critical set. Such a change of variables does not affect the symbol classes $\mathcal{H}^m(k, \ell)$. This concludes the proof of Propositions 3.1 and 3.2.

Now let X be a smooth manifold and Σ a homogeneous isotropic submanifold of $T^*X - 0$. We will denote by $I^m(X, \Sigma)$ the space of all generalized functions, u, such that in every sufficiently small coordinate patch, u can be written as an oscillatory integral of the form (3.4) where a satisfies an asymptotic expansion of the form (3.2). The following is an easy consequence of (3.6).

THEOREM 3.4. *Let* X *and* X_1 *be* n*-dimensional manifolds,* Σ *and* Σ_1 *isotropic submanifolds of* $T^*X - 0$ *and* $T^*X_1 - 0$ *respectively and* $f: T^*X - 0 \to T^*X_1 - 0$ *a homogeneous canonical transformation mapping* Σ *onto* Σ_1. *Let* F *be a Fourier integral operator of order* r *whose underlying canonical transformation is* f. *Then* F *maps* $I^m(X, \Sigma)$ *into* $I^{m+r}(X_1, \Sigma_1)$.

In particular if P is a pseudodifferential operator of order r, P maps $I^m(X, \Sigma)$ into $I^{m+r}(X, \Sigma)$.

§4. THE METAPLECTIC REPRESENTATION

Let $G = Sp(n)$ be the group of linear mappings of R^{2n} leaving fixed the alternating two form $\sum dx_i \wedge dy_i$. As is well known the fundamental group of G is infinite cyclic so there exists a unique connected Lie group $M = Mp(n)$ which double covers G. $Mp(n)$ is called the *metaplectic group.* It has a well-known infinite dimensional unitary representation, the so-called Segal-Shale-Weyl representation, which has certain formal resemblances to the spin representation of the double covering of $SO(n)$. This is constructed as follows: Let n be the $2n+1$ dimensional Heisenberg algebra with basis, $x_1, \cdots, x_n, y_1, y_2, \cdots, y_n, z$, satisfying the bracket relations:

$$[x_i, x_j] = [y_i, y_j] = 0$$

$$(4.1) \qquad [x_i, y_i] = \delta_{ij} z$$

$$[z, \text{anything}] = 0 .$$

Let N be the associated $2n+1$ dimensional Heisenberg group. By the theorem of Stone-von Neumann there exists a unique infinite dimensional unitary representation

$$(4.2) \qquad \rho : N \to \mathcal{U}(H)$$

(H being a separable infinite dimensional Hilbert space) having the property: $\rho(\exp \lambda z) = e^{i\lambda} \mathrm{id}$, $\lambda \in R$. The Hilbert space H can be identified with $L^2(R^n)$. The space of C^∞ vectors* of the representation, ρ,

*The space of C^∞ vectors of a unitary representation $\rho : G \to \mathcal{U}(H)$ of a Lie group, G, is the set of vectors $v \in H$ for which $(\rho(g), v, w)_H$ is a smooth function on G for all $w \in H$. This is the space on which the infinitesimal representation of the Lie algebra of G is defined.

is the space $\mathcal{S}(R^n)$ of rapidly decreasing functions; and the infinitesimal representation of n on $\mathcal{S}$ is described by the formulas

$$(d\rho)(x_i) = \sqrt{-1}\, s_i$$

(4.3) $$(d\rho)(y_i) = \partial/\partial s_i$$

$$(d\rho)(z) = \sqrt{-1}\ \mathrm{id}\ .$$

Now $G = Sp(n)$ acts on $n = R^{2n} \oplus R$ by its usual action on R^{2n} and as the identity on R. It is clear that it preserves the bracket relations (4.1). Therefore, it is a subgroup of the group of automorphisms of n, and also of the group of automorphisms of N.

For every $g \in G$ we get a new irreducible unitary representation of N:

$$\rho_g : N \to \mathfrak{U}(H)$$

defined by $\rho_g(a) = \rho(ga)$. Note that $\rho_g(\exp \lambda z) = \rho(\exp \lambda gz) = \rho(\exp \lambda z) = e^{i\lambda}\mathrm{id}$; so by the Stone-von Neumann theorem ρ_g is unitarily equivalent to ρ. Thus there exists a unitary operator $\mathfrak{U}(g) : H \to H$ such that

$$\mathfrak{U}(g)\rho\mathfrak{U}(g^{-1}) = \rho_g\ .$$

Since ρ is irreducible $\mathfrak{U}(g)$ is determined uniquely except for a multiplicative constant of modulus 1. In particular, there is a function $c : G \times G \to S^1$ such that $\mathfrak{U}(g_1 g_2) = c(g_1, g_2)\mathfrak{U}(g_1)\mathfrak{U}(g_2)$, for all $g_1, g_2 \in G$. Suppose we find a $\tilde{c} : G \to S^1$ such that

(4.4) $$\tilde{c}(g_1)\tilde{c}(g_2)\tilde{c}(g_1 g_2)^{-1} = c(g_1, g_2)\ .$$

Then we can construct a unitary representation of G by setting $\tilde{\mathfrak{U}}(g) = \tilde{c}(g)^{-1}\mathfrak{U}(g)$. It turns out that it is impossible to satisfy the "cocycle condition" (4.4) on G, however it is possible to do so on the double covering of G. Therefore we get a unitary representation $\tau : Mp(n) \to \mathfrak{U}(H)$, the Segal-Shale-Weyl representation mentioned above.

We will need a pretty explicit description of this representation for later on. First of all note that the Lie algebra, $\mathfrak{g}$, of G is the same as the Lie algebra of its double covering and is just the symplectic algebra $sp(n)$. In terms of the basis (x,y) the symplectic algebra can be identified with the space of quadratic forms on R^{2n}. The identification is given by associating to a quadratic form ϕ the linear mapping, L_ϕ, sending (x,y) to $(\bar{x},\bar{y})$ where

$$\begin{aligned} \bar{x} &= \partial\phi/\partial y \\ \bar{y} &= -\partial\phi/\partial x . \end{aligned} \tag{4.5}$$

(Compare with (3.2) above.) Now write

$$\mathfrak{g} = \mathfrak{g}_{-1} \oplus \mathfrak{g}_0 \oplus \mathfrak{g}_{+1} \tag{4.6}$$

where $\mathfrak{g}_{-1}$ corresponds to quadratic forms in x alone, $\mathfrak{g}_0$ to quadratic forms linear in x and in y separately, and $\mathfrak{g}_1$ to quadratic forms in y alone. It is easy to see that the RHS of (4.6) is a *graded* algebra, i.e. $[\mathfrak{g}_{-1}, \mathfrak{g}_{-1}] = 0$, $[\mathfrak{g}_0, \mathfrak{g}_{-1}] \subset \mathfrak{g}_{-1}$, etc.

$\mathfrak{g}_0$ is isomorphic with the Lie algebra, $gL(n)$, the isomorphism being given by $\sum a_{ij}x_i y_j \to (a_{ij})$.

It turns out that the space of C^∞vectors for the representation $\tau : Mp(n) \to \mathcal{U}(H)$ is the same as the space of C^∞ vectors for the representation $\rho : N \to \mathcal{U}(H)$, that is, the Schwartz space $\mathcal{S}(R^n)$. The infinitesimal representation of $\mathfrak{g}$ on $\mathcal{S}$ is given by the following formulas:

A) For $A = L_\phi \in \mathfrak{g}_{-1}$, $\phi = \phi(x)$,

$$(d\tau)(A)f(x) = (1/\sqrt{-1})\phi(x)f(x), \qquad f \in \mathcal{S} . \tag{4.7}$$

B) For $A = L_\psi \in \mathfrak{g}_{+1}$, $\psi = \psi(y)$,

$$(d\tau)(A)f(x) = (1/(2\pi)^n)\int \sqrt{-1}\,\psi(y)e^{ix\cdot y}\hat{f}(y)dy \tag{4.8}$$

$\hat{f}$ being the Fourier transform of f.

C) For $A \in \mathfrak{g}^0$,

$$(d\tau)(A)f = \sum a_{ij}x_i(\partial/\partial x_j)f + (\text{trace } A/2)f \,, \tag{4.9}$$

(a_{ij}) being the matrix representative for A in $g\ell(n)$.

While the metaplectic representation is a little messy to write out on the group level, we can use the formulas above to get a partial picture of it. Let G_{-1}, G_0 and G_1 be the Lie subgroups of G corresponding to $\mathfrak{g}_{-1}$, $\mathfrak{g}_0$ and $\mathfrak{g}_1$. G_{-1} and G_{+1} are just vector groups isomorphic under the exp map with their associated Lie algebras. They therefore have isomorphic copies in the double covering of G which we will continue to denote by G_{-1} and G_{+1}. The action of G_{-1} and G_{+1} on $\mathcal{S}$ is then given by the formulas:

$$\begin{aligned} \exp(L_\phi)f &= e^{-\sqrt{-1}\phi(x)}f \\ \exp(L_\psi)f &= (1/(2\pi)^n)\int e^{\sqrt{-1}\psi(y)}\hat{f}(y)e^{ix\cdot y}\,dy \,. \end{aligned} \tag{4.10}$$

G_0 is just $G\ell(n)$. Its action on $\mathcal{S}$ is a little more complicated to describe. The double covering of $G\ell(n)$ in $Mp(n)$ turns out to have four connected components each isomorphic to $G\ell(n)^+$. This group is called the *metalinear* group and denoted $M\ell(n)$. One of its basic properties is that the determinant function, $\det : G\ell(n) \to R^+$, has an intrinsic square root, $\sqrt{\det}$, when pulled back to $M\ell(n)$. If A is in $M\ell(n)$ and A_0 is its image in $G\ell(n)$ then the action of A on $\mathcal{S}$ is given by the formula

$$Af(x) = f(Ax)\sqrt{\det A} \,. \tag{4.11}$$

This formula plays an important role in what follows. Our first application of it will be to derive a well-known result of Kostant. As above let $n = \mathbf{R}^{2n} \oplus \mathbf{R}$ be the Heisenberg algebra, N the Heisenberg group and $\rho : N \to U(H)$ the Stone-von Neumann representation. n acts on $\mathcal{S}$ by the representation, $d\rho$. By (4.3) $d\rho$ extends to a representation of n on $\mathcal{S}'$, the space of tempered distributions on $\mathbf{R}^n$. Let Λ be a Lagrangian subspace of $\mathbf{R}^{2n}$, let $(\mathrm{Sp})_\Lambda$ be the subgroup of $\mathrm{Sp}(n)$ which maps Λ into itself and let $(\mathrm{Mp})_\Lambda$ be the double covering of $(\mathrm{Sp})_\Lambda$ in $\mathrm{Mp}(n)$. The mapping $(\mathrm{Sp})_\Lambda \to G\ell(\Lambda)$ defined by $A_o \in (\mathrm{Sp})_\Lambda \to A_o|\Lambda$ lifts to a mapping: $(\mathrm{Mp})_\Lambda \to M\ell(\Lambda)$. We will denote by $A|\Lambda$ the image in $M\ell(\Lambda)$ of an element $A \in (\mathrm{Mp})_\Lambda$.

We can think of Λ as an abelian subalgebra of n via the injection $\mathbf{R}^{2n} \to \mathbf{R}^{2n} \oplus \mathbf{R} = n$. We will denote by $\mathcal{S}'_\Lambda$ the space of all $u \in \mathcal{S}'$ with the property

$$d\rho(x)u = 0, \qquad \forall x \in \Lambda . \tag{4.12}$$

If A is in $(\mathrm{Mp})_\Lambda$ then $\tau(A)$ maps $\mathcal{S}'_\Lambda$ into itself.

PROPOSITION 4.1. *$\mathcal{S}'_\Lambda$ is one dimensional. Moreover if $A \in (\mathrm{Mp})_\Lambda$ then for all $u \in \mathcal{S}'_\Lambda$*

$$\tau(A)u = \sqrt{\det (A|\Lambda)}\, u . \tag{4.13}$$

Proof. The symplectic group conjugates any Lagrangian space into any other; so it is enough to prove this for the Lagrangian space $\Lambda = \mathbf{R}^n \oplus \{0\}$. For this space the elements of Λ get represented by the operators "multiplication by x_i" and their linear combinations; and the only tempered distributions on $\mathbf{R}^n$ for which $x_i u = 0$ for all i are the multiples of the delta function. The last assertion is an easy consequence of (4.11). Q.E.D.

To interpret this result we need some new terminology. Let V be a vector space and BV the set of all bases of V. A *metalinear structure* on V is a double covering $\pi : MBV \to BV$ and an action of $M\ell(V)$ on MBV such that if $A \in M\ell(V)$ and A_o is its image in $G\ell(V)$ then $\pi(A\beta) = A_o\pi(\beta)$ for all $\beta \in MBV$. A half-form on V is a map $f : MBV \to C$ such that $f(A\beta) = \sqrt{\det(A)}f(\beta)$ for all $A \in G\ell(V)$. (See [19].) The set of all half-forms on V is a one-dimensional vector space. We will denote it by $\wedge^{1/2}V$. By Proposition 4.1

$$\mathcal{S}'_\Lambda \cong \wedge^{1/2}\Lambda . \tag{4.14}$$

This result is due to Kostant [27]. We will need below an analogue of it for *isotropic* subspaces of $\mathbf{R}^{2n}$. Given a symplectic vector space W let Mp(W) be the metaplectic double covering of Sp(W). Let H(W) be the representation space for the metaplectic representation of Mp(W), let $\mathcal{S}(W)$ be the space of C^∞ vectors in H(W) and let $\mathcal{S}'(W)$ be the topological dual of $\mathcal{S}(W)$. In particular given an isotropic subspace Σ of $\mathbf{R}^{2n}$ let $W = \Sigma^\perp/\Sigma$ and let $\mathcal{S}'_\Sigma$ be the set of all $u \in \mathcal{S}'$ such that

$$d\rho(x)u = 0, \qquad \forall x \in \Sigma .$$

We will show that there is a canonical isomorphism

$$\mathcal{S}'_\Sigma \cong \wedge^{1/2}\Sigma \otimes \mathcal{S}'(W) \tag{4.15}$$

generalizing (4.14).

Proof. Let $(Sp)_\Sigma$ be the subgroup of Sp(n) leaving Σ fixed and let $(Mp)_\Sigma$ be its metaplectic double cover. The canonical maps $(Sp)_\Sigma \to G\ell(\Sigma)$ and $(Sp)_\Sigma \to Sp(W)$ lift to maps: $(Mp)_\Sigma \to M\ell(\Sigma)$ and $(Mp)_\Sigma \to Mp(W)$. To prove (4.15) it is enough to prove it for the special case when Σ is spanned by $x_1, \cdots, x_k$, since any two Σ's of the same dimension are conjugated, one into the other, by Sp(n). For this Σ, $\mathcal{S}'_\Sigma$ consists of all tempered distributions on $\mathbf{R}^n$ such that $x_1u = \cdots = x_ku = 0$. Every such distribution is of the form $\delta_\Sigma \otimes u_W$ where δ_Σ is the delta function on $\mathbf{R}^k$ and u_W a tempered distribution on $\mathbf{R}^{n-k}$. Given $A \in (Mp)_\Sigma$ let A' and A'' be its images in $M\ell(\Sigma)$ and Mp(W) respectively. From (4.10) and (4.11) one easily sees that

$$\tau(A)(\delta_\Sigma \otimes u_W) = \sqrt{\det A'}\,\delta_\Sigma \otimes \tau_W(A'')u_W ,$$

τ_W being the metaplectic representation on $\mathcal{S}'_W$. This establishes the isomorphism (4.15). Q.E.D.

From the conjugate linear pairing of $\mathcal{S}$ with $\mathcal{S}'$ given by the Hilbert space structure on H we get a dualized version of (4.15):

$$\mathcal{S} \to \overline{\wedge}^{-1/2}(\Sigma) \otimes \mathcal{S}_W . \tag{4.16}$$

REMARK. This map will play an essential role in the symbol calculus which we will develop in §6.

We conclude this section by discussing another analogue of (4.15). The symplectic form, ω, on $\mathbf{R}^{2n}$ is also a symplectic form on $\mathbf{R}^{2n} \otimes \mathbf{C}$. Let Λ be a Lagrangian subspace of $\mathbf{R}^{2n} \otimes \mathbf{C}$. Λ is called *positive definite* if the Hermitian form

$$(1/\sqrt{-1})\,\omega(v, \bar{w}), \qquad v, w \in \Lambda ,$$

is positive definite on Λ. Let $\mathcal{S}'_\Lambda$ as above be the set of all $u \in \mathcal{S}'$ such that

$$d\rho(x)u = 0, \qquad \forall x \in \Lambda .$$

PROPOSITION 4.2. *If* Λ *is positive definite then* $\mathcal{S}'_\Lambda$ *is one-dimensional and is contained in* $\mathcal{S}$. *Moreover if* $x \in \mathbf{R}^{2n} \otimes \mathbf{C}$ *satisfies* $d\rho(x)u = 0$ *for all* $u \in \mathcal{S}'_\Lambda$ *then* $x \in \Lambda$.

Proof. It is easy to see that $Sp(n)$ acts transitively on the space of positive definite Lagrangian subspaces of $\mathbf{R}^{2n} \otimes \mathbf{C}$; so it is enough to prove the theorem for the special case

$$\Lambda = \{(x_1 + \sqrt{-1}y_1, x_2 + \sqrt{-1}y_2, \cdots, x_n + \sqrt{-1}y_n)\} .$$

By (4.3) $u \in \mathcal{S}'_\Lambda$ if and only if

$$(\partial/\partial x_i + x_i)u = 0$$

for all i; so u is a constant multiple of $e^{-x^2/2}$. The last assertion we will leave as an exercise for the reader. Q.E.D.

Let U_Λ be the set of all $A \in Mp(n)$ such that the projection of A in $Sp(n)$ preserves Λ. It is clear that if $A \in U_\Lambda$ and $u \in \mathcal{S}'_\Lambda$ then $\tau(A)u = c(A)u$, $c(A)$ being a complex number depending only on A. Since $\tau(A)$ preserves the Hilbert space structure on $\mathcal{S}$, $|c(A)| = 1$. It is also clear that $c(A_1A_2) = c(A_1)c(A_2)$; so c is a unitary character of the group, U_Λ. The following analogue of Proposition 4.1 can be proved by a straightforward computation (which we will omit).

PROPOSITION 4.3. *For all* $A \in U_\Lambda$, $c(A)^2 = \det(A|\Lambda)$.

In §11 we will need:

PROPOSITION 4.4. *Let* A *be in* $Mp(n)$. *Suppose* $\tau(A)$ *maps* $\mathcal{S}'_\Lambda$ *into itself. Then* $A \in U_\Lambda$.

Proof. For $v \in \Lambda$, we have

$$d\rho(Av) = \tau(A)d\rho(v)\tau(A)^{-1}$$

by definition of τ. Therefore if $\tau(A)$ preserves $\mathcal{S}'_\Lambda$, $d\rho(Av)u = 0$ for all $v \in \Lambda$ and all $u \in \mathcal{S}'_\Lambda$. From the second half of Proposition 4.2, we conclude that $A(\Lambda) = \Lambda$. Q.E.D.

An element, e, of $\mathcal{S}$ which is of norm one and lies in some $\mathcal{S}'_\Lambda$ will be called a *vacuum state*. By Proposition 4.1 every vacuum state lies in a unique $\mathcal{S}'_\Lambda$ and the vacuum state in $\mathcal{S}'_\Lambda$ is determined uniquely up to a constant multiple of modulus one.

§5. METALINEAR AND METAPLECTIC STRUCTURES ON MANIFOLDS

We pointed out in §4 that the group $G\ell(n)$ has a natural double covering, the metalinear group $M\ell(n)$. Its fundamental property is that the determinant function lifted to $M\ell(n)$ has an intrinsic square root.

Let X be a smooth manifold, and let BX be the basis bundle of X. Recall that a point of BX is a pair consisting of a point $x \in X$ and a basis $(e_1, \cdots, e_n)$ of T_x. BX is a principal $G\ell(n)$ bundle. By a *metalinear structure* on X we will mean a structure consisting of

a) an $M\ell(n)$ bundle, $MX \to X$

and

b) a double cover $\pi : MX \to BX$ such that $\pi(AB) = \pi(A)\pi(B)$ for all $A \in M\ell$ and $B \in MX$.

It turns out that a metalinear structure exists on X providing a very mild topological condition is satisfied (the vanishing of the square of the first Stiefel-Whitney class) and if X is simply connected this structure is unique. A manifold with a prescribed metalinear structure will from now on be called simply a *metalinear manifold*.

A metalinear manifold possesses a natural line bundle, $\wedge^{1/2}$, called its bundle of $1/2$ forms. Namely consider the representation of $M\ell(n)$ on $\mathbb{C}$ given by $\sqrt{\det}$. $\wedge^{1/2}$ is the line bundle associated to MX by by means of this representation. Note that $\wedge^{1/2} \otimes \wedge^{1/2}$ is the bundle associated to MX by means of the representation "det"; and this is just the determinant bundle or volume form bundle on X. Therefore a section of $\wedge^{1/2}$ is the "square root" of a volume form or, more simply, a *half-form*.

For symplectic manifolds we have a rather analogous situation to that just described. Namely let (Z, Ω) be a symplectic manifold with two form Ω. Let BpZ be the symplectic basis bundle of Z: an element of BpZ is a pair consisting of a point $z \in Z$ and a basis $(v_1, \cdots, v_n, w_1, \cdots, w_n)$ of T_z satisfying $\Omega(v_i, v_j) = \Omega(w_i, w_j) = 0$, $\Omega(v_i, w_j) = \delta_j^i$. BpX is a principal $Sp(n)$ bundle. By a metaplectic structure on X we will mean a structure consisting of

a) an $Mp(n)$ bundle, $MpX \to X$

and

b) a double covering $\pi : MpX \to BpX$ such that $\pi(AB) = \pi(A)\pi(B)$.

Mappings between manifolds, and symplectic mappings between symplectic manifolds have their "meta" counterparts. For example let X and Y be manifolds and $f : X \to Y$ a diffeomorphism. Let f_* be the induced diffeomorphism $f_* : BX \to BY$. If X and Y are metalinear, a metadiffeomorphism will be defined as a pair consisting of a diffeomorphism $f : X \to Y$ and a $M\ell$ diffeomorphism $f_\# : MX \to MY$ covering $f_* : BX \to BY$. Metaplectic mappings between metaplectic manifolds are defined the same way.

There is a rather simple connection between metalinear and metaplectic structures on manifolds, namely:

I. If X is a metalinear manifold there is a canonical metaplectic structure on T^*X.

II. If Z is a metaplectic manifold, every Lagrangian submanifold of Z has a canonical metalinear structure.

These two assertions should not be too surprising considering that the metalinear group sits inside the metaplectic group in simple way. For a proof of I and II, see [2].

We will need below a slightly beefed up version of II. Namely let Σ be an isotropic submanifold of Z. Given $x \in \Sigma$ let Σ_x be the tangent space to Σ at x, and $\Sigma_x^\perp$ its annihilator in T_xZ. Since Σ is isotropic $\Sigma_x^\perp \supset \Sigma_x$ and $\Sigma_x^\perp/\Sigma_x$ is a symplectic vector space of dimension $2(n-k)$, k being the dimension of Σ. The vector bundle E over Σ whose fiber at $x \in \Sigma$ is $E_x = \Sigma_x^\perp/\Sigma_x$ will be called the *symplectic normal bundle* of Σ.

Let $(B \times S)\Sigma$ be the fiber bundle over Σ whose objects are triples consisting of a point $x \in \Sigma$ a basis $(e_1, \cdots, e_k)$ of Σ_x and a symplectic basis $(v_1, \cdots, v_{n-k}, w_1, \cdots, w_{n-k})$ of E_x. This is a principal $G\ell(k) \times Sp(n-k)$ bundle. It turns out that if Z is metaplectic then $(B \times S)\Sigma$ has a four-fold covering by a bundle whose structure group is $M\ell(k) \times Mp(n-k)$.

Consider now the metaplectic representation of $Mp(n-k)$ on the Schwartz space $\mathcal{S}$ and the representation of $M\ell(k)$ on the complex numbers given by $\sqrt{\det}$. The tensor product of these two representations is a representation of $M\ell(k) \times Mp(n-k)$ on $\mathbf{C} \otimes \mathcal{S}$. Let $\mathrm{Spin}(\Sigma)$ be the vector bundle on Σ associated with this representation. $\mathrm{Spin}(\Sigma)$ will be called the *bundle of symplectic spinors associated to* Σ.

The fiber of $\mathrm{Spin}(\Sigma)$ is an infinite dimensional vector space; therefore, a certain amount of caution must be exercised when applying standard vector bundle operations to it. For example how do we define a smooth section of $\mathrm{Spin}(\Sigma)$? Suppose U is a coordinate patch on Σ on which the symplectic normal bundle is trivial. Then a section of $\mathrm{Spin}(\Sigma)$ on U is just a map

$$s : U \to \mathcal{S}(\mathbf{R}^{n-k})$$

which to each $z \in U$ associates a rapidly decreasing function $s(z, x)$ in the variables $x = (x_1, \cdots, x_{n-k})$. We will call s smooth if $s(z, x)$ is simultaneously smooth in both x and z variables.

If $Z = T^*X - 0$, then there is an action of $\mathbf{R}^+$ on Z given by

$$(a, x, \xi) \to (x, a\xi)$$

for $a \in \mathbf{R}^+$, $x \in X$, $\xi \in T_x^*$. To get an induced action of $\mathbf{R}^+$ on $\mathrm{Spin}(\Sigma)$ we first observe that the metaplectic representation can be extended to a representation of $Mp(n) \times \mathbf{R}^+$ by letting $\mathbf{R}^+$ act trivially. Now the action of $\mathbf{R}^+$ on $T^*X - 0$ does not preserve the symplectic structure, but it is conformally symplectic; that is, $a \in \mathbf{R}^+$ maps $T_{(x,\xi)}$ to $T_{(x,a\xi)}$ by a linear mapping which is the composite of a symplectic isomorphism and the conformal mapping $\sqrt{a}\,\mathrm{Id}$. This means that with the convention fixed on above we get an induced action of $\mathbf{R}^+$ on $\mathrm{Spin}(\Sigma)$. We will henceforth denote by $S^m(\Sigma)$ the space of smooth sections of $\mathrm{Spin}(\Sigma)$ which are homogeneous of order m.

§6. ISOTROPIC SUBSPACES OF SYMPLECTIC VECTOR SPACES

In this section we will develop the algebraic tools necessary to formulate our main results in §7. We will begin with some elementary remarks. Let V and W be vector spaces and let Γ be a subspace of $V \oplus W$. We will think of Γ as a collection of pairs, (v, w), with v in V and w in W, i.e. as a "relation" in the set theoretic sense. Given a subspace, Σ, of W we will denote by $\Gamma \circ \Sigma$ the set

$$\{v \in V, \exists w \in \Sigma, (v,w) \in \Gamma\} .$$

(If $f: W \to V$ is a map and $\Gamma = \text{graph } f$, then $\Gamma \circ \Sigma$ is the usual image of Σ with respect to f.) We will denote by Γ^{-1} the space

$$\{(w,v) \in W \oplus V, (v,w) \in \Gamma\} .$$

PROPOSITION 6.1. *Let $\Sigma^{\perp}$ and $\Gamma^{\perp}$ be the annihilator spaces of Σ and Γ in W^* and $V^* \oplus W^*$ respectively. Then $(\Gamma \circ \Sigma)^{\perp} = \Gamma^{\perp} \circ \Sigma^{\perp}$.*

Proof. First suppose that $\Gamma = \text{graph } f$, $f: W \to V$ being a linear map. Let $f^*: V^* \to W^*$ be the transpose of f. Then $u \in \Gamma^{\perp} \circ \Sigma^{\perp} \Longleftrightarrow u \in (f^*)^{-1} \Sigma^{\perp}$ $\Longleftrightarrow < f^*u, w > = 0$, $\forall w \in \Sigma \Longleftrightarrow < u, fw > = 0$, $Aw \in \Sigma \Longleftrightarrow u \in (f \circ \Sigma)^{\perp}$; so in this case the assertion is true. Next suppose $\Gamma = (\text{graph } f)^{-1}$, $f: V \to W$ being a linear mapping. By the previous argument, with Σ replaced by $\Sigma^{\perp}$ and f by f^*, $(f^* \circ \Sigma^{\perp})^{\perp} = f^{-1}(\Sigma)$; so $\Gamma^{\perp} \circ \Sigma^{\perp} = f^* \circ \Sigma^{\perp} = f^{-1}(\Sigma)^{\perp} = (\Gamma \circ \Sigma)^{\perp}$. Finally the general case can be factored into a composition of the two above cases by means of the diagram

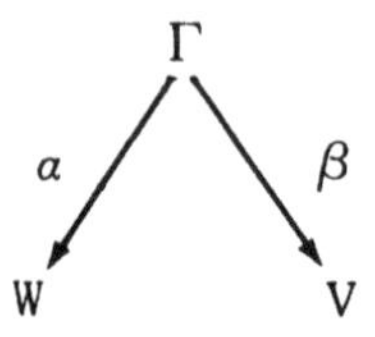

where α and β are the projection of Γ onto V and W respectively and $\Gamma \circ \Sigma = \beta(\alpha^{-1}(\Sigma))$. Q.E.D.

From now on we will suppose that V and W are symplectic vector spaces. Then by means of the symplectic forms on V and W we can identify V with V^*, W with W^* and $V \oplus W$ with $V^* \oplus W^*$. An important special case of the result above is the following

PROPOSITION 6.2. *Let* Γ *be a Lagrangian subspace of* $V \oplus W$. *Then if* Σ *is a subspace of* W, $\Gamma \circ \Sigma^{\perp} = (\Gamma \circ \Sigma)^{\perp}$.

Proof. If Γ is Lagrangian then $\Gamma = \Gamma^{\perp}$ in Proposition 6.1. Q.E.D.

COROLLARY. *If* Σ *is isotropic,* $\Gamma \circ \Sigma$ *is isotropic and if* Σ *is Lagrangian* $\Gamma \circ \Sigma$ *is Lagrangian.*

Assume, as above, that Γ is a Lagrangian subspace of $V \oplus W$. Let α be the projection of $V \oplus W$ onto W.

PROPOSITION 6.3. $\alpha(\Gamma)$ *is a co-isotropic subspace of* W *and*

$$\alpha(\Gamma)^{\perp} = \{w \in W, (0, w) \in \Gamma\}. \tag{6.1}$$

Proof. $\alpha(\Gamma) = \Gamma^{-1} \circ V$; so $\alpha(\Gamma)^{\perp} = \Gamma^{-1} \circ \{0\}$ by Proposition 6.2. Q.E.D.

Let Σ be an isotropic subspace of W and set

$$U_0 = \alpha(\Gamma)^{\perp} \cap \Sigma = \{w \in \Sigma, (0, w) \in \Gamma\} \tag{6.2}$$

and

$$U_1 = \alpha(\Gamma)^{\perp} \cap \Sigma^{\perp} = \{w \in \Sigma^{\perp}, (0, w) \in \Gamma\}. \tag{6.3}$$

U_0 and U_1 are isotropic subspace of W, both contained in $\Sigma^{\perp}$, and

$U_0 = U_1 \cap \Sigma$. Let U be the image of U_1 in $\Sigma^\perp/\Sigma$. U is an isotropic subspace of $\Sigma^\perp/\Sigma$ isomorphic to U_1/U_0.

PROPOSITION 6.4. *There is a canonical isomorphism*

$$(\Gamma\circ\Sigma)^\perp/(\Gamma\circ\Sigma) \cong U^\perp/U . \tag{6.4}$$

Proof. $(U_1+\Sigma)^\perp = a(\Gamma)\cap\Sigma^\perp+\Sigma$; so there is a canonical surjective map

$$a(\Gamma)\cap\Sigma^\perp \to U^\perp . \tag{6.5}$$

Given $u \in U^\perp$, choose $(v,w)\in\Gamma$ such that $w\in\Sigma^\perp$ and the image of w in $\Sigma^\perp/\Sigma$ is u. (This is possible by (6.5).) Then $v \in \Gamma\circ\Sigma^\perp$. We leave it as an easy exercise to show that the correspondence which associates v to u defines a map $\Phi : U^\perp/U \to (\Gamma\circ\Sigma)^\perp/\Gamma\circ\Sigma$. In the other direction, given $v\in(\Gamma\circ\Sigma)^\perp$ there exists a $w\in\Sigma^\perp$ such that $(v,w)\in\Gamma$ and $w\in\Sigma^\perp$. Thus $w\in a(\Gamma)\cap\Sigma^\perp$. Let u be the image of w in $U^\perp$ under the map (6.5). Again we leave it as an easy exercise to show that the correspondence which associates u to v extends to a map $\Psi:(\Gamma\circ\Sigma)^\perp/(\Gamma\circ\Sigma)\to U^\perp/U$. Finally it is trivial to see that Φ and Ψ are inverses of each other. Q.E.D.

Given a symplectic vector space and an isotropic subspace, Σ, let $Mp(\Sigma^\perp/\Sigma)\to U(H)$ be the metaplectic representation of $Mp(\Sigma^\perp/\Sigma)$ and let $\mathcal{S}(\Sigma^\perp/\Sigma)$ be the space of C^∞ vectors in H. We will denote by $\mathrm{Spin}(\Sigma)$ the tensor product

$$\mathrm{Spin}(\Sigma) = \wedge^{1/2}(\Sigma)\otimes\mathcal{S}(\Sigma^\perp/\Sigma) . \tag{6.6}$$

PROPOSITION 6.5. *There is a canonical* $\mathbf{R}$*-linear map*:

$$\mathrm{Spin}(\Sigma)\otimes\wedge^{1/2}(\Gamma) \to \mathrm{Spin}(\Gamma\circ\Sigma)\otimes|\wedge|(U_0) . \tag{6.7}$$

Proof. Let ρ be the map: $\Gamma\oplus\Sigma\to W$, $((v,w),w_1)\to w-w_1$. By (6.3), Image $\rho = U_1^\perp$. The kernel of ρ is the set of all pairs $(v,w)\in\Gamma$ such that $w\in\Sigma$. Thus there is a surjective map

$$\ker\rho \to \Gamma\circ\Sigma .$$

The kernel of this map is the set of all $w \in \Sigma$ such that $(0,w) \in \Gamma$. By (6.2) this is just U_o. Therefore we get a pair of exact sequences

$$0 \to \ker \rho \to \Gamma \oplus \Sigma \to U_1 \to 0 \tag{6.8}$$

and

$$0 \to U_o \to \ker \rho \to \Gamma \circ \Sigma \to 0 . \tag{6.9}$$

From the symplectic structure on W we get an isomorphism

$$U_1^{\perp} \cong (W/U_1)^* \tag{6.10}$$

and finally by (4.16) we get a map

$$\mathcal{S}(\Sigma^{\perp}/\Sigma) \to \overline{\wedge}^{-1/2}(U) \otimes \mathcal{S}(U^{\perp}/U) . \tag{6.11}$$

Using (6.8), (6.9) and (6.10) we can identify certain spaces of half-forms. From (6.10) we get

$$\wedge^{1/2}(U_1^{\perp}) \cong \wedge^{-1/2}(W) \otimes \wedge^{1/2}(U_1) .$$

Since W is a symplectic space it has a canonical non-zero volume form, hence a canonical half-form; so $\wedge^{-1/2}(W) \cong \mathbb{C}$ and $\wedge^{1/2}(U_1^{\perp}) \cong \wedge^{1/2}(U_1)$. Combining this with complex conjugation we get an $\mathbb{R}$-linear isomorphism

$$\overline{\wedge}^{-1/2}(U_1) \cong \wedge^{-1/2}(U_1^{\perp}) . \tag{6.12}$$

From (6.8) and (6.9) we get

$$\wedge^{-1/2}(U_1^{\perp}) \cong \wedge^{1/2}(U_o) \otimes \wedge^{1/2}(\Gamma \circ \Sigma) \otimes \wedge^{-1/2}(\Gamma) \otimes \wedge^{-1/2}(\Sigma) . \tag{6.13}$$

By definition $\mathrm{Spin}(\Sigma^{\perp}/\Sigma) = \wedge^{1/2}(\Sigma) \otimes \mathcal{S}(\Sigma^{\perp}/\Sigma)$. Since $U = U_1/U_o$ and $U^{\perp}/U = (\Gamma \circ \Sigma)^{\perp}/(\Gamma \circ \Sigma)$ we get from (6.11) a map of $\mathrm{Spin}(\Sigma^{\perp}/\Sigma) \otimes \wedge^{1/2}(\Gamma)$ into

$$\wedge^{1/2}(\Sigma) \otimes \wedge^{1/2}(\Gamma) \otimes \overline{\wedge}^{1/2}(U_o) \otimes \overline{\wedge}^{-1/2}(U_1) \otimes \mathcal{S}((\Gamma \circ \Sigma)^{\perp}/(\Gamma \circ \Sigma)) . \tag{6.14}$$

Substituting (6.12) and (6.13) we get (after some cancellation) a canonical isomorphism of (6.14) with

$$|\wedge|(U_0) \otimes \{\wedge^{1/2}(\Gamma \circ \Sigma) \otimes \mathcal{S}((\Gamma \circ \Sigma)^{\perp}/(\Gamma \circ \Sigma)\} .$$

However the expression in brackets in just $\mathrm{Spin}(\Gamma \circ \Sigma)$. Q.E.D.

§7. THE COMPOSITION THEOREM

Given manifolds X, Y and Z and maps $f: X \to Z$ and $g: Y \to Z$, f and g are said to *intersect cleanly* if their fiber product, F,

$$\begin{array}{ccc} F & \longrightarrow & X \\ \big\downarrow & & \big\downarrow f \\ Y & \xrightarrow{\;g\;} & Z \end{array} \tag{7.1}$$

is a submanifold of $X \times Y$ and in addition for each $p \in F$, $p = (x, y)$,

(7.2)

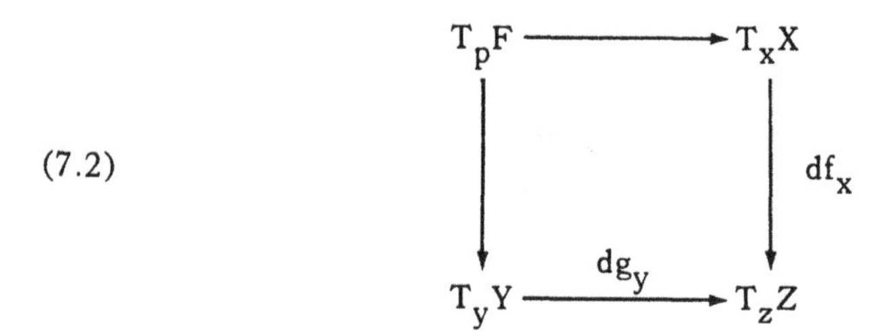

is a fiber product diagram. The non-negative integer

$$e = \dim F + \dim Z - (\dim X + \dim Y) \tag{7.3}$$

is called the *excess* of the clean intersection (7.1).

Let X and Y be manifolds, and Γ and Σ closed homogeneous submanifolds of $T^*(X \times Y) - 0$ and $T^*Y - 0$ respectively, Γ being Lagrangian and Σ isotropic. Let $\Gamma \circ \Sigma$ be the set of (x, ζ) in T^*X for which there exists (y, η) in Σ with (x, ζ, y, η) in Γ. We will make the following assumptions:

a) Γ contains no vectors of the form $(x, \zeta, y, 0)$.

b) $\Gamma \circ \Sigma$ contains no zero vectors.

c) Let Γ' be the projection of Γ in $X \times Y$. Then $\Gamma' \to X$ is proper.

d) The following is a clean fiber product:

(7.4)

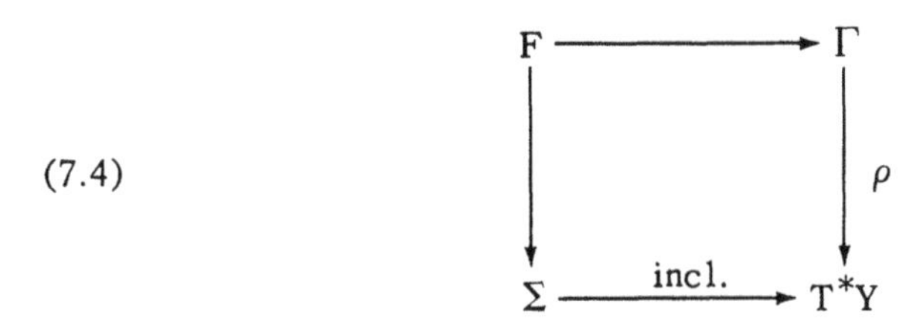

e) The map $\tau : F \to T^*X$ defined by the composite of the top arrow in d) and $\pi : \Gamma \to T^*X$ is of constant rank.

PROPOSITION 7.1. *If (7.4) holds then* $\Gamma \circ \Sigma$ *is an (immersed) isotropic submanifold of* $T^*X - 0$ *and the map* $\tau : F \to \Gamma \circ \Sigma$ *is a fiber mapping with compact fiber.*

Proof. The first part of the proposition follows immediately from Proposition 6.1. The fiber of $\tau : F \to \Gamma \circ \Sigma$ must be compact for otherwise in view of the homogeneity of Γ and Σ and in view of c), b) would be violated. Q.E.D.

Condition d) in (7.4) can be formulated somewhat differently. Let p be a point of F and let q and r be its images in T^*Y and T^*X respectively. Let F_p be the tangent space to F at p, Γ_p the tangent space to Γ at p, Σ_q the tangent space to Σ at q, W_q the tangent space to T^*Y at q and V_r the tangent space to T^*X at r. Γ_r is a Lagrangian subspace of $V_r \oplus W_q$ and Σ_q an isotropic subspace of W_q; so we are in the situation considered in §6. Identifying F_p with the set

(7.5) $$\{(v, w) \in \Gamma, w \in \Sigma_q\}$$

the kernel of the map $(d\tau)_p : F_p \to V_r$ gets identified with the set

(7.6) $$(U_o)_p = \{w \in \Sigma_q, (0,w) \in \Gamma_q\}.$$

(Compare with 6.2.) As in §6 set

(7.7) $$(U_1)_p = \{w \in \Sigma_q^{\perp}, (0,w) \in \Gamma_p\}$$

and

(7.8) $$U_p = (U_1)_p/(U_o)_p .$$

PROPOSITION 7.2. *If condition d) of (7.4) is satisfied then the dimension of the space (7.7) is independent of* p.

Proof. Consider the map $\Gamma_p \oplus \Sigma_q \to W_q$, $((v,w_1),w_2) \to w_1 - w_2$. By Proposition 6.3 the image of this map is the symplectic orthocomplement of the space (7.7), and the kernel is just the space (7.5). By d) of (7.4), (7.5) is the tangent space to F at p; so its dimension is independent of p. Q.E.D.

COROLLARY. *Condition e) of (7.4) holds if and only if* $\dim U_p$ *is independent of* p.

Proof. Condition e) is equivalent to the condition that $\dim(U_o)_p$ be independent of p. Q.E.D.

We will assume from now on that all the conditions (4.4) hold. We will denote by $U \to F$ the vector bundle whose fiber at p is U_p. We will call U the excess *bundle* of the pair (Γ, Σ).

Suppose now that X and Y are metalinear manifolds and hence that their cotangent bundles are metaplectic manifolds. (See §5.) Let Ω_Γ be a fixed half-form on Γ homogeneous of degree r. We will show that there is a canonical R-linear map depending on Ω_Γ

(7.9) $$S^m(\Sigma) \to S^{m+r-\frac{1}{2}(\dim Y - e)}(\Gamma \circ \Sigma)$$

where e is the excess in the diagram (7.4) d). Let σ be an element of $S^m(\Sigma)$. Let p be a point of F and let r and q be its images in T^*X

and T^*Y. By (6.7) there is a canonical map

$$\text{Spin}(\Sigma_q) \otimes \wedge^{1/2}(\Gamma_p) \to |\wedge|(U_o)_p \otimes \text{Spin}(\Gamma \circ \Sigma)_r . \tag{7.10}$$

Consider the image of $\sigma_q \otimes (\Omega_\Gamma)_p$ under this map. By (7.6) $(U_o)_p$ is the tangent space to the fiber of the fiber mapping $\tau : F \to \Gamma \circ \Sigma$. Thus, on the fiber above r, we have an object which is an element of $\text{Spin}(\Gamma \circ \Sigma)_r$ tensored with a density along the fiber. By Proposition 7.1, the fiber is compact; so we can integrate the density over the fiber and we are left with an element $\sigma'(r)$ of $\text{Spin}(\Gamma \circ \Sigma)_r$. It is clear that σ' depends smoothly on r; so all we have left to check is that it has the right degree of homogeneity. To do so we must show that the map (7.10) has the right degree of homogeneity as p varies along a ray in F. Going back to the proof of Proposition 6.5, consider the map 6.10:

$$(U_1^\perp)_p \xrightarrow{\cong} (W/U_1)_p^* . \tag{7.11}$$

This map maps $w \in (U_1)_p^\perp$ into $w \lrcorner \Omega_q$, where Ω is the symplectic form on T^*Y. Since Ω is homogeneous of degree one, (7.11) is homogeneous of degree one. Therefore, the induced map on $\frac{1}{2}$ forms:

$$\wedge^{1/2}(U_1^\perp)_p \to \wedge^{1/2}(W/U_1)_p^* = \wedge^{1/2}(U_1) \otimes \wedge^{-1/2}(W_q) \tag{7.12}$$

is homogeneous of degree, $-\frac{1}{2}\dim(U_1^\perp)_p$. The volume form, Ω^n, $n = \dim Y$, is homogeneous of degree n; so the trivializing map $\wedge^{-1/2}(W_q) \to \mathbb{C}$ sending $\Omega^{-n/2}$ into 1 is homogeneous of degree $n/2$.

Composing this with the map (7.12) we get a map

$$\wedge^{1/2}(U_1^\perp)_p \to \wedge^{1/2}(U_1) \tag{7.13}$$

which is homogeneous of degree:

$$-\frac{1}{2}\dim U_1^\perp + \frac{\dim Y}{2} .$$

In the proof of Proposition 7.2 we showed that $\dim U_1 = e$; so $\dim U_1^\perp =$

$\dim W - e = 2\dim Y - e$. Thus the degree of homogeneity of (7.13) is $-\frac{1}{2}(\dim Y - e)$. From this one easily sees that the map (7.10) has the same degree of homogeneity. (Compare (6.12) and (6.14).) Q.E.D.

As a first application of (7.9) we will define symbols for the distributions, $I^m(X, \Sigma)$, of §3.

We recall that a generalized function, u, on X belongs to the space $I^m(X, \Sigma)$ if it is locally expressible by an oscillatory integral of the form

$$a(x, \tau, n/\sqrt{\tau})\, e^{i\phi(x,\tau,\eta)}\, d\tau\, d\eta \tag{7.14}$$

with $\phi : X \times \mathbf{R}^N \to \mathbf{R}$ a defining phase function of Σ and $a \in \mathcal{H}^{m-N/2}(k,\ell)$, $N = k+\ell$. (See (3.4).) To attach symbols to these distributions we will have to make some additional assumptions about (7.14):

(I) The amplitude in (7.14) admits an asymptotic expansion of the form (3.2).

(II) The order of homogeneity of the terms in (3.2) are either integer or half-integer.

(III) X is a metalinear manifold and u is a generalized *half-form*, i.e. u is locally expressible by an oscillatory integral of the form

$$\left(\int a(x, \tau, \eta/\sqrt{\tau})\, e^{i\phi(x,\tau,\eta)}\, d\tau\, d\eta\right)\sqrt{dx} \tag{7.14$'$}$$

the term in parentheses being as before.

We equip the projection map $\pi : X \times \mathbf{R}^N \to X$ with the structure of a morphism of half-forms by requiring that $\pi^*\sqrt{dx} = \sqrt{dx\, d\tau\, d\eta}$. Let Γ be the conormal bundle of graph π in $T^*(X \times X \times \mathbf{R}^n)$. Then the pull-back map, π^*, on half-forms is a Fourier integral operator associated with the canonical relation, Γ. Its symbol is a half-form Ω_Γ on Γ. If we identify Γ with π^*T^*X and use coordinates (x,ξ,θ) on Γ where (x,ξ) are the cotangent coordinates on T^*X and $\theta=(\tau,\eta)$ then

$$\Omega_\Gamma = \sqrt{dx\,d\xi\,d\theta}\,.$$

Let $d\phi : X\times \mathbf{R}^N \to T^*(X\times\mathbf{R}^N)$ be the map, $(x,\theta)\to(d\phi)_{x,\theta}$, and let Σ_ϕ be the image under $d\phi$ of the set $\eta_1=\cdots=\eta_\ell=0$. Then Σ_ϕ is an isotropic submanifold of $T^*(X\times\mathbf{R}^N)$. It is not hard to see that Γ and Σ_ϕ intersect transversally in $T^*(X\times\mathbf{R}^N)$ and that $\Sigma=\Gamma\circ\Sigma_\phi$. (The second statement is just another way of saying that ϕ is a defining phase function for Σ.) Consider now on $\Sigma_\phi\cap(|\tau|=1)$ the symplectic spinor

$$\nu = \sqrt{dx\,d\tau}\otimes a_{m_0}(x,\tau,\eta)\sqrt{d\eta} \tag{7.15}$$

where $a_{m_0}(x,\tau,\eta)$ is the leading term in the asymptotic expansion (3.2). Extend ν to a symplectic spinor on all of Σ_ϕ by requiring it to have degree of homogeneity $m-N/2$.

DEFINITION 7.3. We will define the symbol, $\sigma(u)$, of the distribution (7.14)′ to be the image of ν under the mapping (7.9). It is easy to check that $\sigma(u)\in S^m(\Sigma)$.

PROPOSITION 7.4. *The definition of* $\sigma(u)$ *is independent of the choice of the parametrization, (7.14)′ hence the correspondence,* $u\to\sigma(u)$, *extends to a canonical, globally defined mapping*

$$\sigma : I^m(X,\Sigma) \to S^m(\Sigma) \tag{7.16}$$

whose kernel is $I^{m-1/2}(X,\Sigma)$.

This theorem was proved in §8 of [15]. In [15] we used a slightly different method of defining the symbol from that used above. The two definitions can be reconciled by an argument very similar to that of §3. We will not bother to do so here.

We can now formulate what is in some sense the main result of this monograph. Let X and Y be metalinear manifolds, let Σ be a homogeneous, isotropic submanifold of T^*Y-0 and let Γ be a homogeneous Lagrangian submanifold of $T^*(X\times Y)-0$. Assume Γ and Σ satisfy the axioms (7.4) a)-e). Let $\mathfrak{k}$ be a generalized half-form on $X\times Y$ belonging to the class of Lagrangian distributions $I^r(X\times Y,\Gamma)$. (See [25].) From $\mathfrak{k}$ we get a continuous $\mathbf{R}$-linear operator, K, which maps smooth compactly supported half-forms on Y into generalized half-forms on X by the formula

$$Ku = \int_Y \mathfrak{k}(x,y)\bar{u}(y)\,. \tag{7.17}$$

Note that $\bar{u}$ is a conjugate half-form on Y; so the integrand of the expression on the left can be interpreted as a half-form on X times a density in Y. This justifies our integrating it over Y. It follows from Hörmander [25], Theorem 2.5.14, that (7.17) makes sense even when u is a generalized half-form providing the set

$$WF(\mathfrak{k})\circ WF(u)$$

contains no zero vectors.

THEOREM 7.5. K *maps* $I^s(Y,\Sigma)$ *into* $I^{r+s-\frac{1}{2}(\dim Y-e)}(\Gamma\circ\Sigma)$. *Moreover, for* $u\in I^s(Y,\Sigma)$

$$\sigma(Ku) = \sigma(K)\circ\sigma(u) \tag{7.18}$$

$\sigma(K)$ *being a half-form on* Γ *(the symbol of the Lagrangian distribution,* $\mathfrak{k}$*) and the right-hand side of (4.9) being the image of* $\sigma(u)$ *with respect to the map (7.9).*

REMARK. For Σ Lagrangian this theorem was proved in [12], §4.

§8. THE PROOF OF THEOREM 7.5

We will make some simplifying assumptions about Γ and Σ. To justify these assumptions we will need

LEMMA 8.1. *Let* X *and* X_1 *be* n*-dimensional manifolds,* Σ *and* Σ_1 *homogeneous isotropic submanifolds, of* T^*X-0 *and* T^*X_1-0, p *a point of* Σ, p_1 *a point of* Σ_1 *and* $f:(\Sigma,p)\to(\Sigma_1,p_1)$ *a germ of homogeneous diffeomorphism. Then there exists a germ of homogeneous canonical transformation* $g:(T^*X,p)\to(T^*X_1,p_1)$ *extending* f.

LEMMA 8.2. *Let* Σ *be a homogeneous isotropic submanifold of* T^*X-0, S *a homogeneous submanifold of* Σ *and* $U\to S$ *an isotropic subbundle of* $N|S$ *where* $N_s=(T_s\Sigma)^{\perp}/T_s\Sigma$. *Assume* U *is invariant under the action of the homotheties* $(x,\xi)\to(x,\lambda\xi)$ *on* N. *Then for every* $s_o\in S$ *there exists a neighborhood,* $\mathcal{O}'$, *of* s_o *in* Σ *and a homogeneous isotropic submanifold,* Σ', *of* T^*X-0 *such that* $\Sigma'\supset\Sigma\cap\mathcal{O}'$ *and for all* $s\in S\cap\mathcal{O}'$, $T_s\Sigma'/T_s\Sigma=U_s$.

For the proof of Lemma 8.1 see §3 of [15]. To prove Lemma 8.2, let $\mathcal{O}$ be a neighborhood of s_o in T^*X-0 and f a function on $\mathcal{O}$ with the following properties

(i) $df_{s_o}\neq 0$.

(ii) $f=0$ on $\Sigma\cap\mathcal{O}$.

(iii) f is homogeneous of degree one.

(iv) For all $s\in S\cap\mathcal{O}$ the image of $H_f(s)$ in $(T_s\Sigma)^{\perp}/T_s\Sigma$ lies in E_s, H_f being the Hamiltonian vector field associated with f.

The existence of f is an easy consequence of the Whitney extension theorem. Let Σ_1, S_1 and U_1^o be the saturations of Σ, S and U by the flow, $t \to \exp tH_f$, and let U_1 be the image of U_1^o in the symplectic conormal bundle of Σ_1. Argue by induction with Σ_1, S_1, U_1 in place of Σ, S, U. Q.E.D.

As a first step in proving Theorem 7.5 we will show that we can, without loss of generality, make some additional restrictions on Σ and Γ besides the conditions (7.4). First of all we will show that we can assume $\Gamma \to T^*Y - 0$ is an *imbedding*. To see this suppose we are given arbitrary Γ and Σ satisfying the conditions (7.4). If $u \in I^s(Y, \Sigma)$ consider the tensor product, $u \otimes k$, as a generalized half-form on $Y \times Y \times X$. We will show below (§9) that this belongs to $I^{r+s}(Y \times Y \times X, \Sigma \times \Gamma)$ near the diagonal in $T^*Y \times T^*Y \times T^*X$. Let $\Delta : Y \times X \to Y \times Y \times X$ be the diagonal map and $\pi : Y \times X \to X$ the trivial fibering. Then $Ku = \pi_* \Delta^*(u \otimes k)$. The operator $\pi_* \Delta^*$ is a Fourier integral operator from $C_o^\infty(Y \times Y \times X)$ into $C_o^\infty(X)$ associated with the canonical relation

$$\Gamma_1 = \{((y, \eta, y, -\eta, x, \xi), (x, -\xi)), ((x, \xi) \in T^*X - 0, (y, \eta) \in T^*Y - 0)\} .$$

Letting $\Sigma_1 = \Sigma \times \Gamma$, Γ_1 and Σ_1 satisfy all the hypotheses (7.4) and in addition $\Gamma_1 \to T^*Y_1$ is injective, with $Y_1 = Y \times Y \times X$.

From now on we will assume Γ, in Theorem 7.5, is an imbedded submanifold of $T^*Y - 0$. By Proposition 6.3 this imbedded submanifold is co-*isotropic*. By condition d) of (7.4) Γ and Σ intersect cleanly in a submanifold, F, of Σ. For each $p \in F$, let U_p be the image of $(T_p\Gamma)^\perp \cap (T_p\Sigma)^\perp$ in $(T_p\Sigma)^\perp / T_p\Sigma$. In §7 we showed that the assignment $p \to U_p$ defines a vector bundle $U \to F$, called the excess bundle. We will now show that both Σ and U can be chosen to have very simple forms.

As in §2 we will write $\mathbf{R}^n = \mathbf{R}^k \times \mathbf{R}^\ell$ with variables (t, y) and dual variables (τ, η). By Lemmas 8.1 and 8.2 we can find a germ of homogeneous canonical transformation $f : (T^*Y, p) \to T^*\mathbf{R}^n$ mapping Σ onto the isotropic manifold $t = y = \eta = 0$ and mapping the vector subbundle U of the symplectic conormal bundle of Σ into a bundle over $f(F)$ which is spanned at each point by $\partial/\partial y_1, \cdots, \partial/\partial y_r$. If K_1 is an F.I.O., associated with the canonical transformation f, which is elliptic near p then, replacing u by $K_1 u$ and K by $K \circ K_1^t$ in Theorem 4.1, we are reduced to proving Theorem 7.5 with the additional hypotheses

(8.1)

a) The manifold, Y, is $\mathbf{R}^n$ and Σ is the isotropic manifold, $0 = t = y = \eta$, in $T^*\mathbf{R}^n$.

b) $\Gamma \to T^*\mathbf{R}^n$ is an imbedding and Γ and Σ intersect cleanly in a submanifold F of Σ.

c) The excess bundle, $U \to F$, is the following bundle: If $p \in F$, the fiber of U_p is just the image of the vectors

$$\partial/\partial y_1, \partial/\partial y_2, \cdots, \partial/\partial y_r \quad \text{in} \quad (T_p\Sigma)^{\perp}/T_p\Sigma .$$

We will now see what these conditions mean when we parametrize Γ locally by means of a phase function.

Let $\phi = \phi(x,t,y,\theta)$ be a smooth function on $X \times \mathbf{R}^k \times \mathbf{R}^\ell \times (\mathbf{R}^N - 0)$ which is homogeneous of degree one in θ and is non-degenerate in the θ variables in the sense of Hörmander [25]. Let $C_\phi = \{(x,t,y,\theta),\ \partial\phi/\partial\theta_1 = \cdots = \partial\phi/\partial\theta_N = 0\}$ be the critical set of ϕ and let Γ be the diffeomorphic image of C_ϕ under the map

$$(8.2) \qquad (x,t,y,\theta) \to (x, \partial\phi/\partial x, t, \partial\phi/\partial t, y, \partial\phi/\partial y) .$$

(Naturally all of Γ cannot be parametrized this way, but we are interested for the moment only in the local situation about a given point, $\gamma_0 \in \Gamma$.) The submanifold, F, of Γ gets mapped by the inverse of (8.2) onto a submanifold of C_ϕ which we will denote by S. S is defined by the equations

$$(8.3) \qquad \begin{aligned} &\partial\phi/\partial\theta_i = 0, \ i = 1, \cdots, N \quad \partial\phi/\partial y_i = y_i = 0, \ i = 1, \cdots, \ell \\ &t_i = 0, \ i = 1, \cdots, k . \end{aligned}$$

Let p be a fixed point on F and s the corresponding point on S. If $w \in U_p$ then by (7.6) and (7.7) there exists $v \in T_s C_\phi$ such that

$$d(x, \partial\phi/\partial x)_s v = 0$$

and

$$d(t, y, \partial\phi/\partial t, \partial\phi/\partial y)_s v = w \bmod T_p\Sigma .$$

By $(8.1)_c$ $\partial/\partial y_i$, $0 \leq i \leq r$ projects onto an element of U_p; so there exists a $v_i \in T_s C_\phi$ such that

$$d(x, \partial\phi/\partial x)_s v_i = 0$$

and

(8.4) $$d(t, y, \partial\phi/\partial t, \partial\phi/\partial y)_s v_i = \partial/\partial y_i \bmod T_s\Sigma = \partial/\partial y_i + \sum b_{i\alpha}\partial/\partial\tau_\alpha .$$

v_i is not unique; however by $(7.4)_e$ the dimension of the space (7.6) does not change from point to point; so we can choose v_i to depend smoothly on $s \in S$. Since the projection of v_i on $X \times R^k \times R^\ell$ is $\partial/\partial y_i$, v_i must have the form

(8.5) $$v_i = \partial/\partial y_i + \sum a_{ij}(\partial/\partial\theta_j)$$

where the a_{ij}'s are smooth functions on S. We will show that by making a change of coordinates in the θ variables, leaving the (x,t,y) variables fixed, we get rid of the a_{ij}'s in (8.5). First we will prove:

LEMMA 8.3. *Given $s_o \in S$ we can find an open set, $\mathcal{O}$, containing s_o in $X \times R^k \times R^\ell \times R^N - 0$ and smooth functions a_{ij} on $\mathcal{O}$, homogeneous of order 1 such that the a_{ij}'s extend the a_{ij}'s of (8.5) and such that the extended v_i's satisfy $[v_i, v_j] = 0$ for $i, j \leq r$.*

Proof. By (5.3) S is contained in the set $S' : t = y = 0$. First of all extend the a_{ij}'s arbitrarily to homogeneous functions of degree 1 on this set. We will continue to denote the extended vector fields by v_i. For each $x \in X$ let $M_{\bar{x}}$ be the manifold: $x = \bar{x}$, $y_{r+1} = \cdots = y_\ell = 0$, $t = 0$. $M_{\bar{x}}$ intersects S' in the set of points $\{x, \theta, 0, 0\}$. At these points the space spanned by the v_i's is tangent to $M_{\bar{x}}$ and transverse to $M_{\bar{x}} \cap S'$. It is clear that we can foliate $M_{\bar{x}}$ in such a way that the foliation is r

dimensional and the tangent space to the foliation is spanned by the v_i's at points of $M_{\bar{x}} \cap S'$. It is also clear that we can make this foliation depend smoothly on $\bar{x}$. Finally we can construct r dimensional foliations of the sets $x = \bar{x}$, $t = \bar{t}$, $y_i = \bar{y}_i$, $i = r+1, \cdots, \ell$, such that these foliations depend smoothly on the parameters $\bar{x}, \bar{t}, \bar{y}$ and extend the foliations on the $M_{\bar{x}}$'s. (All of this construction is in some small neighborhood of s_o.) It is not hard to show that this construction can be made to be invariant under the homotheties, $\theta \to \lambda\theta$, giving us a homogeneous r-dimensional foliation of all of $X \times R^n \times (R^N - 0)$. For every point sufficiently close to s_o, there is a unique vector v_i of the form $\partial/\partial y_i + \sum a_{ij}(\partial/\partial\theta_j)$ tangent to this foliation. Consider the Lie bracket, $[v_i, v_j]$. On the one hand it is a linear combination of the v_k's since the v_i's are tangent to the foliation. On the other hand, since the v_i's are of the form (8.5), it is a linear combination of the $\partial/\partial\theta_i$'s. Hence $[v_i, v_j] = 0$.

Q.E.D.

COROLLARY. *Given* $s_o \in S$ *there exists a conic neighborhood,* U, *of* s_o *in* $X \times R^n \times R^N - 0$ *and a homogeneous diffeomorphism,* χ, *such that*

(8.6)

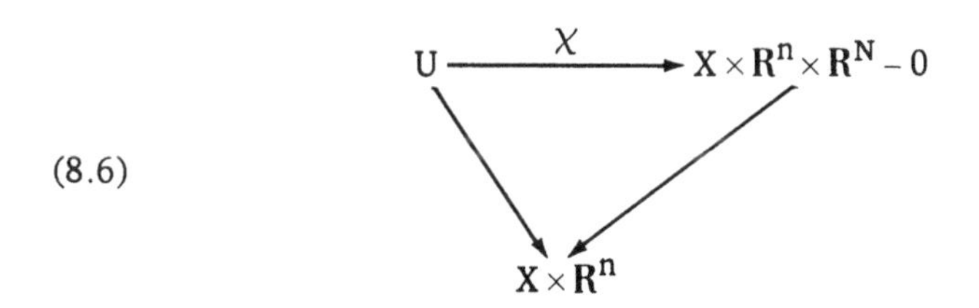

commutes and $(d\chi)v_i = \partial/\partial y_i$, $i = 1, \cdots, r$, *where* v_i *is the extension of* (8.5) *constructed above.*

Proof. If we impose the additional requirement that χ is the identity map on the set $y_1 = \cdots = y_r = 0$, then the conditions above determine χ completely since χ intertwines the action of R^r on $X \times R^N \times R^N - 0$ given by

$$(x, t, y, \theta) \to (x, t, y_1 + a_1, \cdots, y_r + a_r, y_{r+1}, \cdots, y_k, \theta)$$

and the action given by

$$(x,t,y,\theta) \to (\exp a_1 v_1)(\exp a_2 v_2)\cdots(\exp a_r v_r)(x,y,t,\theta)\,.$$

Q.E.D.

Replacing the phase function ϕ by the phase function $(\chi^{-1})^*\phi$, which is also a parametrizing phase function for Γ, we can henceforth assume that the a_{ij}'s in (8.5) are identically zero. This considerably simplifies the equations (8.4). These equations now take the form:

$$\begin{aligned} &\partial\phi/\partial y_i \partial y_j = 0 && i = 1,\cdots,r,\ \ j = 1,\cdots,k \\ &\partial\phi/\partial y_i \partial \theta_j = 0 && i = 1,\cdots,r,\ \ j = 1,\cdots,N \\ &\partial\phi/\partial y_i \partial x_j = 0 && i = 1,\cdots,r,\ \ j = 1,\cdots,\dim X \end{aligned} \tag{8.7}$$

on the set S. Note that, by (8.3), $\partial\phi/\partial y_i = 0$ on S. Note also that by (8.3) the functions $t_1,\cdots,t_k$ vanish on S. We can summarize the content of (8.7):

LEMMA 8.4. *Let* $b_{ij} = \partial\phi/\partial y_i \partial t_j(x,y,0,\theta)$, $i = 1,\cdots,r$, $j = 1,\cdots,k$. *Then*

$$\partial\phi/\partial y_i - \sum_{j=1}^{k} b_{ij} t_j \tag{8.8}$$

vanishes to second order on S.

Now let us come back to the distribution, Ku, in Theorem 7.5. The Schwartz kernel of K belongs to the space $I^r(X \times R^n, \Gamma)$; so we can write it locally as an oscillatory integral of the form

$$\int k(x,t,y,\theta)\, e^{i\phi(x,t,y,\theta)}\, d\theta$$

ϕ being the phase function above and k being a classical symbol of order $m_1 = r - N/2$. Since u belongs to the space $I^s(R^n, \Sigma)$ we can write it as an oscillatory integral of the form

$$u = \int a(\tau, \eta/\sqrt{\tau})\, e^{i(y\eta + t\tau)}\, d\eta\, d\tau \ ,$$

$a(\tau, \eta)$ being of order $m_s = s - n/2$ in τ.

Thus, for Ku we get the expression

$$(8.9) \qquad (Ku)(x) = \int \ell\, e^{i(\phi + y''\eta'' + t\tau)} a(\tau, \eta/\sqrt{\tau})\, e^{iy'\eta'} dy\, dt\, d\eta\, d\tau\, d\theta \ .$$

Here we have written y' for $(y_1, \cdots, y_r)$ and y'' for $(y_{r+1}, \cdots, y_\ell)$ and ditto for η' and η''. We will rewrite (8.9) in the form

$$\int |\tau|^{r/2}\, \ell\, e^{i(\phi + t\tau + y''\eta'')} \left(\int a(\tau, \eta', \eta''/\sqrt{\tau})\, e^{i\sqrt{\tau}(y'\eta')} dy'\, d\eta' \right) dm$$

where $dm = dy''\, dt\, d\eta''\, d\tau\, d\theta$. We can formally simplify the expression in parentheses by stationary phase. This gives us (formally) the expansion

(8.10)

$$\frac{1}{(2\pi)^r} \sum_p \int |\tau|^{-p/2} \frac{i^p}{p!} \left(\sum \frac{\partial}{\partial y'} \frac{\partial}{\partial \eta'} \right)^p \{\ell\, e^{i(\phi + y''\eta'' + t\tau)} a(\tau, \eta', \eta''/\sqrt{\tau})\}\Big|_{y' = \eta' = 0} dm \, .$$

At first glance it might appear hopeless to make any analytic sense of this expansion. Indeed, if we perform the indicated differentiations in y' we get terms arising of the following sort:

$$(8.11)_p \qquad \int |\tau|^{(-p/2)} \frac{i^p}{p!} \left(\frac{\partial \phi}{\partial y'} \right)^{\nu} \ell\, e^{i(\phi + y''\eta'' + t\tau)} a^{(\nu)}(\tau, 0, \eta''/\sqrt{\tau})\, dm$$

and

$$(8.12)_p \qquad \int |\tau|^{-p} \frac{(-1)^p}{(2p)!} \left(\frac{\partial^2 \phi}{\partial y' \partial y'} \right)^{\nu} \ell\, e^{i(\phi + y''\eta'' + t\tau)} a^{(\nu')}(\tau, 0, \eta''/\sqrt{\tau})\, dm$$

where $|\nu| = p$, $|\nu'| = 2p$ and $a^{(\nu)}(\tau, \eta', \eta'') = \left(\frac{\partial}{\partial \eta'} \right)^{\nu} a(\tau, \eta', \eta'')$. We also get, of course, mixed terms involving first and second derivatives in y

and terms involving third and higher order derivatives. The degree of homogeneity of the amplitude in $(8.11)_p$ is $p/2 + m_1 + m_2$ and of the amplitude in $(8.12)_p$ $m_1 + m_2$; so in $(8.12)_p$ the order is non-decreasing as p tends to infinity and in $(8.11)_p$ the order is even tending to infinity! Using (8.8), however, we will be able to rewrite these terms in such a way that the asymptotic expansion (8.10) does make sense. First let us analyze the term (8.11). Using (8.8) this can be written as a sum of terms of the following sort:

$$(8.13)_a \qquad \int |\tau|^{-p/2} A(x,y'',t,\theta)\, t^{\alpha} e^{i(\phi+y''\eta''+t\tau)} a(\tau,\eta''/\sqrt{\tau})\, dm$$

$$(8.13)_b \qquad \int |\tau|^{-p/2} B(x,y'',t,\theta)\left(\frac{\partial\phi}{\partial\theta}\right)^{\beta} e^{i(\phi+y''\eta''+t\tau)} a(\tau,\eta''/\sqrt{\tau})\, dm$$

$$(8.13)_c \qquad \int |\tau|^{-p/2} C(x,y'',t,\theta)\left(\frac{\partial\phi}{\partial y''}\right)^{\gamma} e^{i(\phi+y''\eta''+t\tau)} a(\tau,\eta''/\sqrt{\tau})\, dm$$

and

$$(8.13)_d \qquad \int |\tau|^{-p/2} D(x,y'',t,\theta)\,(y'')^{\delta} e^{i(\phi+y''\eta''+t\tau)} a(\tau,\eta''/\sqrt{\tau})\, dm ,$$

plus mixed terms of the same sort. To simplify we have written $a(\tau,\eta'')$ for $a^{(\nu)}(\tau,0,\eta'')$ in $(8.11)_p$. In these expressions $|\alpha| = p$, $|\beta| = |\gamma| = |\delta| = 2p$, A, B and D are of order $m_1 + p$ in θ and C is of order $m_1 - p$. By an integration by parts we can rewrite $(8.13)_a$ in the form

$$A e^{i(\phi+y''\eta''+t\tau)} D_{\tau}^{\alpha}\{|\tau|^{-p/2} a(\tau,\eta''/\sqrt{\tau})\}\, dm ;$$

so $(8.13)_a$ is actually of order $m_1 + m_2 - p/2$. Since $\frac{\partial\phi}{\partial\theta_i} e^{i(\phi+y''\eta''+t\tau)} = \frac{1}{\sqrt{-1}}\frac{\partial}{\partial\theta_i} e^{i(\phi+y''\eta''+t\tau)}$, and $|\beta| = 2p$ we can perform p integrations by parts in $(8.13)_b$ and reduce it to an integral of the form

$$\int |\tau|^{-p/2} B'(x,y'',t,\theta) e^{i(\phi+y''\eta''+t\tau)} dm$$

where B' is of order m_1. Hence $(8.13)_b$ is actually of order $m_1+m_2-p/2$ as well. To simplify $(8.13)_c$ we will set $(\partial\phi/\partial y'')^\gamma = (\partial\phi/\partial y''+\eta''-\eta'')^\gamma$ and expand in powers of $((\partial\phi/\partial y'')+\eta'')$ and η''. Then $(8.13)_c$ becomes a sum of the terms

$$(8.14) \quad \int |\tau|^{-p/2} C(x,y'',t,\theta)\left(\frac{\partial\phi}{\partial y''}+\eta''\right)^\gamma e^{i(\phi+y''\eta''+t\tau)} a(\tau,\eta''/\sqrt{\tau})dm$$

$$(8.15) \quad \int |\tau|^{-p/2} C(x,y'',t,\theta)(\eta'')^\gamma e^{i(\phi+y''\eta''+t\tau)} a(\tau,\eta''/\sqrt{\tau})dm$$

and terms involving mixed powers of $((\partial\phi/\partial y'')+\eta'')$ and η''. Since $((\partial\phi/\partial y'')+\eta'')e^{i(\phi+y''\eta''+t\tau)} = (1/\sqrt{-1})(\partial/\partial y'')e^{i(\phi+y''\eta''+t\tau)}$, (8.14) can be reduced to a sum of terms of order $\le m_1+m_2-p/2$ by integration by parts just as in $(8.13)_p$. As for (8.15), we can rewrite it in the form

$$\int |\tau|^{p/2} C(x,y,t,\theta) e^{i(\phi+y''\eta''+t\tau)}(\eta''/\sqrt{\tau})^\gamma a(\tau,\eta''/\sqrt{\tau})dm$$

which is again of order $m_1+m_2-p/2$. Similar remarks apply to the mixed terms.

Finally $(8.13)_d$ can be rewritten as

$$\int |\tau|^{-p/2}(i^p/p!)D(x,y'',t,\theta)(D^\delta_{\eta''}e^{i(\phi+y''\eta''+t\tau)})a(\tau,\eta''/\sqrt{\tau})dm$$

which by an integration by parts becomes

$$\int |\tau|^{-3p/2}(i^p/p!)D(x,y'',t,\theta)e^{i(\phi+y''\eta''+t\tau)}a^{(\delta)}(\tau,\eta''/\sqrt{\tau})dm$$

where $a^{(\delta)}(\tau,\eta'') = D^{\delta}_{\eta''} a(\tau,\eta'')$. This shows that $(8.13)_d$ is also of order $m_1 + m_2 - p/2$. We let the reader check that the mixed terms which one gets, (in addition to $(8.13)_a$ - $(8.13)_d$) when one expands out the integrand of (8.11) using (8.8), are also all of order $m_1 + m_2 - p/2$.

We still have to contend with terms like $(8.12)_p$ and terms involving higher derivatives of y' as well. By (8.7) $\partial^2\phi/\partial y'\partial y'$ vanishes on S, so we can write it as a linear combination of the functions (8.2) and reduce the integrand of $(8.12)_p$ to order $m_1 + m_2 - p/2$ or lower by integrations by parts just as above. The contributions coming from third or higher order derivatives of y' in (8.10) will be, for instance, terms like

$$\int |\tau|^{-3p/2} (\partial^3\phi/\partial y_1^3)^p e^{i(\phi + y''\eta'' + t\tau)} ((\partial^{3p}/\partial\eta_1^{3p}) a)(\tau, \eta''/\sqrt{\tau})\, dm$$

which are already of sufficiently low order that we can ignore them. This proves that the orders of the terms in (8.10) are going to $-\infty$ as p gets large. We still have to show that (8.10) is, in fact, an asymptotic expansion of (8.9). The proof of this is essentially a recapitulation of the arguments above, applied to the remainder term in the partial stationary phase expansion of (8.9). We will omit details.

We must now analyze the individual terms in the asymptotic expansion (8.10). Before we do so it will be helpful to make some general remarks about oscillatory integrals of the form (7.14) with *clean* phase functions. For the moment let X be an open subset of $\mathbf{R}^n$, let $\mathbf{R}^N = \mathbf{R}^k \times \mathbf{R}^\ell$ with coordinates $\theta = (\tau,\eta)$ and let $\phi : X \times \mathbf{R}^n - 0 \to \mathbf{R}$ be a smooth function which is homogeneous in θ of degree one. We will say that ϕ is a *clean* phase function relative to the set, $\eta = 0$, if the following is true

(8.16) i) The set

$$Z = \{(x,\theta), \partial\phi/\partial\theta = 0, \eta = 0\}$$

is a submanifold of $X \times \mathbf{R}^N$ of codimension m.

ii) At each point of Z exactly m of the differentials $d(\partial\phi/\partial\theta)$, $d\eta$ are independent.

iii) Each of the differentials $d\eta_i$ is a linear combination of dx, $d(\partial\phi/\partial x)$ and $d(\partial\phi/\partial\tau)$.

LEMMA 8.5. *If ϕ satisfies (8.16) i), ii) and iii) the map $(x, \partial\phi/\partial x): Z \to T^*X$ is of constant rank everywhere and its image is an $(n-\ell)$-dimensional homogeneous, isotropic submanifold, Σ, of $T^*X - 0$.*

Proof. We can apply the results of §7 to the pair Γ_o, Σ_o where Σ_o is the intersection of graph $d\phi$ in $T^*(X \times \mathbf{R}^N - 0)$ with the set, $\eta = 0$, and Γ_o is the conormal bundle of graph π in $T^*(X \times X \times \mathbf{R}^N - 0)$ where $\pi: X \times \mathbf{R}^N - 0 \to X$ is the trivial projection. Condition (ii) is the clean intersection condition $(7.4)_d$ for this pair and condition (iii) is equivalent to the condition, that, for each point $p \in Z$, the space U_p of (7.8) is zero; which implies $(7.4)_e$. To conclude we observe that $\Gamma_o \circ \Sigma_o$ is equal to the image of $(x, \partial\phi/\partial x): Z \to T^*X - 0$. Q.E.D.

REMARK. Condition (iii) implies that the differentials $d(\partial\phi/\partial\eta_1), \cdots, d(\partial\phi/\partial\eta_\ell)$, $d\eta_1, \cdots, d\eta_\ell$ are linearly independent on Z. Indeed suppose that at some point of Z

$$\sum a_i d(\partial\phi/\partial\eta_i) + b_i d\eta_i = 0 . \tag{8.17}$$

Let v be the vector, $\sum a_i(\partial/\partial\eta_i)$. Evaluating (8.17) on the vectors $\partial/\partial x$ and $\partial/\partial\tau$ we get

$$d(\partial\phi/\partial x)v = d(\partial\phi/\partial\tau)v = 0 ;$$

so v is annihilated by dx, $d(\partial\phi/\partial x)$ and $d(\partial\phi/\partial\tau)$. Hence by (iii) v is annihilated by $d\eta$; so $a_1 = a_2 = \cdots = a_\ell = 0$ in (8.17). Since the $d\eta$'s are linearly independent, $b_1 = \cdots = b_\ell = 0$ as well.

Now consider oscillatory integrals of the form

$$u = \int a(x, \tau, \eta/\sqrt{\tau})e^{i\phi(x,\theta)}d\theta \tag{8.18}$$

where $a \in \mathcal{H}^{m-N/2}(k, \ell)$ and ϕ is a clean phase function.

PROPOSITION 8.6. *u is in $I^{m+e/2}(X, \Sigma)$ where $e = \dim Z - \dim X + k$.*

Proof. Given $z_0 \in Z$ we can choose certain of the τ variables, say $\tau_1, \cdots, \tau_r$ such that

$$d\eta, d(\partial\phi/\partial\eta) \quad \text{and} \quad d(\partial\phi/\partial\tau_1), \cdots, d(\partial\phi/\partial\tau_r)$$

are linearly independent at each point of Z near z_0 and such that, at each such point,

$$d(\partial\phi/\partial\tau_{r+1}), \cdots, d(\partial\phi/\partial\tau_k)$$

are linear combinations of them. Here $k - r = e$. We will now rewrite (8.18) in such a way that the τ_i's, $i > r$ occur as harmless parameters. Let $\tau = (\tau', \tau'')$ where $\tau' = (\tau_1, \cdots, \tau_r)$ and $\tau'' = (\tau_{r+1}, \cdots, \tau_\ell)$. Without loss of generality we can assume $|\tau''| \le |\tau'|$ on Z. By homogeneity

$$\phi(x, \tau', \tau'', \eta) = |\tau'|\phi(x, \tau'/|\tau'|, \omega, (\eta/|\tau'|)) = \phi'(x, \tau', \omega, \eta)$$

where ω is $\tau''/|\tau'|$ and ϕ' is a new phase function homogeneous of degree one in τ' and η. We can rewrite (8.18) as

$$\left(\int a'(x, \tau', \omega, \eta/\sqrt{\tau'})e^{i\phi'(x,\tau',\omega,\eta)} d\tau' d\eta\right) d\omega \tag{8.19}$$

where $a'(x, \tau', \omega, \eta) = |\tau'|^e a(x, \tau', |\tau'|\omega, (1/\sqrt{1+\omega^2})\eta)$. Let u_ω be the inner integral in (8.19). We will show that for each ω, u_ω belongs to $I^{m+3/2}(X, \Sigma)$ and depends smoothly on ω. By (8.19) this will show that u is in $I^{m+e/2}(X, \Sigma)$. It is easy to see that if $d(\partial\phi/\partial\tau')$ and $d(\partial\phi/\partial\eta)$ are linearly independent, then so are $d(\partial\phi'/\partial\tau')$ and $d(\partial\phi'/\partial\eta)$. Therefore, for fixed ω, $\phi': X \times \mathbf{R}^{N-e} - 0 \to \mathbf{R}$ is a non-degenerate phase function in the sense of §3 and parametrizes an $n-k$ dimensional isotropic submanifold of T^*X. However, $(x, \partial\phi'/\partial x)(x, \tau', \omega, \eta) = (x, \partial\phi/\partial x)(x, \tau', \tau'', \eta)$; so this submanifold sits inside Σ. Therefore it must be identical with Σ by Lemma 8.5. Finally a' is of order $m - N/2 + e$ and there are $N-e$ phase variables in the inner integral in (8.19) so $u_\omega \in I^{m+e/2}(X, \Sigma)$.

Q.E.D.

We will now apply these results to each of the terms in the expansion (8.10). A typical term in this expansion is of the form

$$\int k(x,t,y'',\theta) e^{i(\phi(x,t,y'',\theta)+y''\eta''+t\tau)} a(\tau, \eta''/\sqrt{\tau}) dm \tag{8.20}$$

where k is a classical symbol in the x, t, y'', θ variables, a is in $\mathcal{H}^m(k, \ell)$, $\phi(x,t,y'',\theta) = \phi(x,t,y,\theta)|_{y'=0}$ and $dm = dt\,dy''d\theta\,d\tau\,d\eta''$. We would like to regard (8.20) as a damped oscillatory integral in the phase variables $(\theta, y'', \eta'', t, \tau)$. One minor difficulty with doing so is that the variables y'' and t are not homogeneous. However, this difficulty is not serious since y'' and t are interchangeable with $|\theta|y''$ and $|\theta|t$ which are homogeneous. Another difficulty is that the amplitude in (8.20) is not quite in appropriate form. However, we can write this amplitude as $\tilde{a}(x, \zeta, \eta/\sqrt{\zeta})$ where $\zeta = (\theta, |\theta|y'', |\theta|t, \tau)$ and $a(x,\zeta,\eta) = |\theta|^{r-k-\ell}(x,y'',t,\theta)$ $a(\tau, (\sqrt{\zeta}/\sqrt{\tau})\eta'')$, and (8.20) now becomes

$$\int \tilde{a}(x,\zeta,\eta''/\sqrt{\zeta}) e^{i\tilde{\phi}(x,\zeta,\eta'')} d\zeta\, d\eta'' \tag{8.21}$$

with $\tilde{\phi}(x,\zeta,\eta'') = \phi(x,y'',t,\theta) + y''\eta'' + t\tau$ where $\tilde{a}$ and $\tilde{\phi}$ have the same properties as the amplitude and phase in (8.18). To apply Proposition 8.6 to (8.21) we need to show

LEMMA 8.7. *The phase function, $\tilde{\phi}$, is clean with respect to the set* $\eta = 0$.

Proof. The critical set of $\tilde{\phi}$ intersects $\eta' = 0$ in the set

(8.22)
$$\begin{aligned} \partial\phi/\partial\theta &= 0 \\ \partial\phi/\partial y'' &= 0 \\ (\partial\phi/\partial t) + \tau &= 0 \\ t &= 0 \\ y'' &= 0\,. \end{aligned}$$

By (8.3) these are the defining equations of the set S. Since S is diffeomorphic to the intersection of Γ and Σ the fact that (8.22) is a "clean" system of equations is equivalent to the hypothesis $(7.4)_d$. Next we have to check (8.16)(iii): we have to show that at all points, s, of S the system of linear equations in T_s:

(8.23)
$$\begin{aligned} 0 = dx = dt &= d(\partial\phi/\partial x) = d((\partial\phi/\partial y'')-\eta'') \\ &= d((\partial\phi/\partial t)-\tau) = d(\partial\phi/\partial\theta) \end{aligned}$$

implies $d\eta'' = 0$. Suppose there existed a vector

$$v = a(\partial/\partial y'') + b(\partial/\partial\theta) + d(\partial/\partial\tau) + d(\partial/\partial\eta'')$$

annihilated by all the differentials (8.23). Let

$$w = a(\partial/\partial y'') + b(\partial/\partial\theta)\,.$$

Then from the equations above we get

$$d(x, \partial\phi/\partial x)w = 0$$

$$d(y'', , \partial\phi/\partial y'', \partial\phi/\partial t)w = a(\partial/\partial y'') + c(\partial/\partial\tau) + d(\partial/\partial\eta'')\,.$$

This shows that $a(\partial/\partial y'') + c(\partial/\partial\tau) + d(\partial/\partial\eta'')$ is in the space (7.7) at s. However, by assumption, this space is spanned by the $\partial/\partial y''$'s mod $T_s\Sigma$; so $b = c = 0$ and, in particular, v is annihilated by $d\eta''$. Q.E.D.

We will say a few words about the symbol of the distribution (8.9). Its symbol is the same as the symbol of the leading term in (8.10):

$$1/(2\pi)^r \int k(x,t,y,\theta)\, a(\tau,\eta',\eta''/\sqrt{\tau})\, e^{(\phi(x,t,y,\theta)+y''\eta''+t\tau)}\Big|_{y'=\eta'=0} dm$$

$\mathcal{k}$ and a being the same $\mathcal{k}$ and a as in (8.9). On the symplectic spinor level the map $a(\tau, \eta', \eta'') \to a(\tau, 0, \eta'')$ is the map (6.7). By (8.19) the integral (8.21) is a superposition of oscillatory integrals with non-degenerate phase functions, u_ω, each of whose symbols is computable by the procedure outlined in §7. The symbol of (8.21) is the integral over ω of the symbols of the u_ω's. It is not hard to see that the intrinsic meaning of this integral is the fiber integral over the fibration $F \to \Gamma \circ \Sigma$ described in (7.4).

§9. PULL-BACKS, PUSH-FORWARDS AND EXTERIOR TENSOR PRODUCTS

Let X and Y be manifolds with preassigned metalinear structures. Let $f : X \to Y$ be a smooth map. We will assume f is equipped with the structure of a morphism of half-forms in the sense of [18], Chapter VI, §5. This allows us to define the pull-back, f^*u, of a half-form u on Y. Also, if f is a fiber mapping it allows us to define the push-forward or "fiber integral", f_*u, of a compactly supported half-form on X. Let Γ be the conormal bundle of the graph of f in $T^*(X \times Y)$. The following are immediate corollaries of Theorem 7.5.

THEOREM 9.1. *Let* Σ *be a homogeneous, isotropic submanifold of* $T^*Y - 0$ *such that* Σ *and* Γ *satisfy (7.4). Then if* $u \in I^r(\Sigma)$, $f^*u \in I^{r+e/2}(f^*\Sigma)$.

THEOREM 9.2. *Let* f *be a fiber mapping. Let* Σ *be a homogeneous, isotropic submanifold of* $T^*X - 0$ *such that* Γ *and* Σ *satisfy (7.4). Then if* $u \in I^r(\Sigma)$, $f_*u \in I^{r-N/2+e/2}(f_*\Sigma)$, N *being the fiber dimension of* f.

REMARK. We can also write down explicit formulas for the symbols of f^*u and f_*u in terms of the symbol of u just as in [18], Chapter VI, §5. Unfortunately, however, these formulas are not much simpler than the general formula described in §7.

Let X_1 and X_2 be metalinear manifolds and u_i, $i = 1, 2$, a generalized half-form on X_i. The exterior tensor product $u_1 \boxtimes u_2$ is a generalized half-form on $X_1 \times X_2$.

THEOREM 9.3. *Suppose* u_i *is in* $I^{r_i}(\Sigma_i)$. *Let* $\mathcal{C}$ *be a conic neighborhood of the set* $\Sigma_1\times 0 \cup 0\times\Sigma_2$ *in* $T^*(X_1\times X_2)$. *Then*

$$u_1 \boxtimes u_2 = \alpha + \beta \tag{9.1}$$

where $\alpha \in I^{r_1+r_2}(\Sigma_1\times\Sigma_2)$ *and the wave front set of* β *is in* $\mathcal{C}$. *In addition*

$$\sigma(u_1 \boxtimes u_2) = \sigma(u_1)\boxtimes\sigma(u_2) \tag{9.2}$$

on $\Sigma_1\times\Sigma_2 - \mathcal{C}$.

REMARK. (9.2) requires a little explanation. If V_i, $i = 1,2$ is a metaplectic vector space and H_i its space of symplectic spinors then $H_1 \boxtimes H_2$ can be viewed as a subspace of the space of symplectic spinors for $V_1\times V_2$. Thus the right-hand side of (9.2) can be viewed as an element of $\mathrm{Spin}(\Sigma_1\times\Sigma_2)$.

Proof. Because of invariance with respect to conjugation by invertible Fourier integral operators, it is enough to prove the theorem above when Σ_i is a linear subspace of the cotangent space to the origin in $\mathbf{R}^{n_i}$. Let cotangent coordinates $\zeta_i = (\tau_i, \eta_i)$ be chosen so that Σ_i is defined by the equation $\eta_i = 0$. Then, modulo smooth functions,

$$u_i = \int f_i(\tau_i, \eta_i/\sqrt{|\tau_i|})e^{\sqrt{-1}\,x_i\zeta_i}\,d\zeta_i\,, \qquad i = 1,2\,,$$

where $f_i(\tau_i,\eta_i)$ is rapidly decreasing in η_i, is zero for $|\tau_i|$ small and admits an asymptotic expansion in homogeneous functions of τ_i for τ_i large. (See §3.) Thus

$$u_1 \boxtimes u_2 = \int f_1(\tau_1, \eta_1/\sqrt{|\tau_1|})f_2(\tau_2, \eta_2/\sqrt{|\tau_2|})e^{ix\zeta}\,d\zeta$$

where $x = (x_1, x_2)$ and $\zeta = (\zeta_1, \zeta_2)$. Let ρ be a homogeneous function of τ of degree 0 which is zero near $\Sigma_1 \times 0 \cup 0 \times \Sigma_2$ and is one on the set

$$\varepsilon|\tau_1| < |\tau_2| < (1/\varepsilon)|\tau_1| . \tag{9.3}$$

Set

$$f(\tau,\eta) = \rho(\tau) f_1(\tau_1, \sqrt{|\tau|}\,\eta_1/\sqrt{|\tau_1|}) f_2(\tau_2, \sqrt{|\tau|}\,\eta_2/\sqrt{|\tau_2|}) .$$

Then $u_1 \boxtimes u_2 = a + \beta$ where

$$a = \int f(\tau, \eta/\sqrt{|\tau|}) e^{ix\zeta} d\zeta .$$

a belongs to $I^{r_1+r_2}(\Sigma_1 \times \Sigma_2)$, and $WF(\beta) \subseteq \mathcal{C}$ providing we choose ε small enough in (9.3). It is clear from (9.4) that $\sigma(a) = \sigma(u_1) \boxtimes \sigma(u_2)$ on $\Sigma_1 \times \Sigma_2 - \mathcal{C}$. Q.E.D.

It may happen that a complicated operation on distributions can be factored into a sequence of push-forwards, pull-backs, and exterior tensor products. If so, we can apply the theorems above to see how the $I^r(\Sigma)$'s are affected. We will discuss two examples:

EXAMPLE 1. (An operator applied to a generalized function.) We will prove a generalization of Theorem 7.5 in which Γ is isotropic rather than Lagrangian.

THEOREM 9.4. *Let Γ and Σ be closed homogeneous isotropic submanifolds of $T^*(X \times Y) - 0$ and $T^*Y - 0$ respectively, satisfying the conditions (7.4). Let k be in $I^r(X \times Y, \Gamma)$ and let K be the operator defined by (7.17). Then K maps $I^s(\Sigma)$ into $I^{r+s-1/2(\dim Y - e)}(\Gamma \circ \Sigma)$.*

Proof. Let u be in $I^s(\Sigma)$. Then $Ku = \pi_* \Delta^* k \boxtimes \bar{u}$, where $\Delta : X \times Y \to X \times X \times Y$ is the diagonal map and π is the projection of $X \times Y$ onto X. Choose $\mathcal{C}$ in Theorem 9.3 to be the subset of $T^*(X \times Y \times Y)$ consisting of all triples $(x, \zeta, y_1, \eta_1, y_2, \eta_2)$ for which

(9.6) $$|\eta_1| + |\eta_2| \neq 0, \quad |\eta_1| < \frac{1}{2}|\eta_2| \quad \text{or} \quad |\eta_2| < \frac{1}{2}|\eta_1| .$$

By (7.4) a) this is an open neighborhood of $\Gamma \times 0 \cup 0 \times \Sigma$. $\Delta^*\mathcal{C}$ consists of all pairs $(x, \zeta, y, \eta_1+\eta_2)$ for which η_1 and η_2 satisfy (9.6). For such a pair $\eta_1+\eta_2 \neq 0$ so $\pi_*\Delta^*\mathcal{C}$ is empty. Therefore, if β is a generalized function on $X \times Y \times Y$ with its wave front set contained in $\mathcal{C}$, $\pi_*\Delta^*\beta$ is smooth by the results of Hörmander on wave front sets in [25], §2.5. We now write $k \boxtimes \bar{u}$ as a sum of the form (9.1). As we have just observed, $\pi_*\Delta^*\beta$ is smooth. Since u is in $I^s(\Sigma)$, $\bar{u}$ is in $I^s(-\Sigma)$ and $\pi_*\Delta_*(\Gamma \times (-\Sigma)) = \Gamma \circ \Sigma$. By successive application of Theorems 9.1, 9.2, and 9.3, one sees that $\pi_*\Delta^*a$ is in $I^{r+s-1/2(\dim Y - e)}(\Gamma \circ \Sigma)$.*

EXAMPLE 2. (The composition of two operators.) Let Γ_1 and Γ_2 be closed homogeneous isotropic submanifolds of $T^*(X \times Y) - 0$ and $T^*(Y \times Z) - 0$ respectively. We will assume Γ_1 contains no vectors of the form $(x, \zeta, y, 0)$ or $(x, 0, y, \eta)$, and we will make the same assumption about Γ_2. We will also assume that the projection of Γ_1 (Γ_2) on $X \times Y$ is a properly supported subset of $X \times Y$ $(Y \times Z)$. Finally we will make two assumptions about how Γ_1 and Γ_2 are situated relative to one another.

FIRST ASSUMPTION. The following diagram is clean.

(9.7)

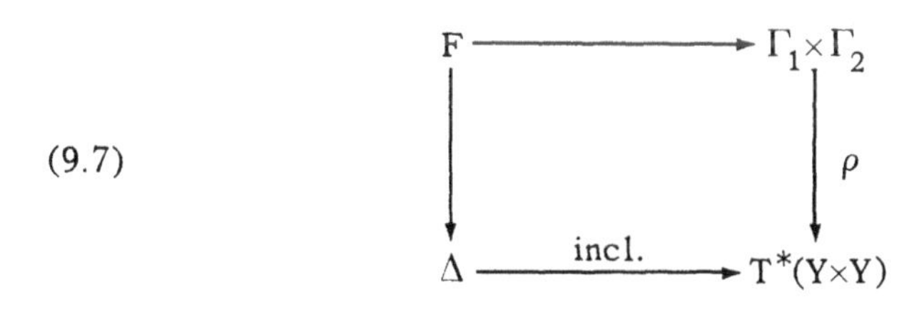

Δ being the diagonal in $T^*Y \times T^*Y$ and ρ the projection of $T^*(X \times Y \times T \times Z)$ onto $T^*(Y \times Y)$.

---

*We have glossed over a slight difficulty. The fact that Γ and Σ satisfy (7.4) does not insure that π and Δ will *separately* satisfy the transversality (or cleanness) hypothesis of Theorems 9.1 and 9.2. The way out of this difficulty is to apply Theorem 7.5 directly to $\pi_*\Delta^*$.

SECOND ASSUMPTION. The map $\tau: F \to T^*(X \times Z)$ given by the composition of the top arrow in (6.7) with projection of $\Gamma_1 \times \Gamma_2$ into $T^*(X \times Z)$ is of constant rank.

These assumptions imply that $\Gamma_1 \circ \Gamma_2$ is an immersed isotropic submanifold of $T^*(X \times Z)$. Let e be the excess in diagram (9.7).

THEOREM 9.5. *Let* $k_i \in I^{r_i}(\Gamma_i)$ *and let* K_i *be the operator (7.17) associated with* k_i, $i = 1, 2$. *Let* k *be the Schwartz kernel of* $K_1 \circ K_2$. *Then*

$$k \in I^{r_1 + r_2 - 1/2(\dim Y - e)}(\Gamma_1 \times \Gamma_2).$$

Proof. The proof is virtually the same as before. The main observation to make is that $k = \pi_* \Delta^* k_1 \boxtimes k_2$ where $\Delta: X \times Y \times Z \to X \times Y \times Y \times Z$ is the diagonal map and $\pi: X \times Y \times Z \to X \times Z$ the projection map. Q.E.D.

An important subcase of this example is the following. Let Σ be a closed homogeneous submanifold $T^*X - 0$. Let $\Sigma^\#$ be the set

$$\{(x, \zeta, x, -\zeta), (x, \zeta) \in \Sigma\}$$

in $T^*(X \times X)$. This is clearly an isotropic submanifold. (It sits in the conormal bundle of the diagonal.) Let

$$\mathcal{H}^m(\Sigma) = I^{m+1/2(\dim X - d)}(\Sigma^\#) \tag{9.8}$$

where d is the codimension of Σ. Let $OP\mathcal{H}^m(\Sigma)$ be the space of operators having elements of $\mathcal{H}^m(\Sigma)$ as kernels. These are *Hermite operators* in the sense of [4].

THEOREM 9.6. *If* $K_i \in OP\mathcal{H}^{m_i}(\Sigma)$, $i = 1, 2$, *then* $K_1 \circ K_2$ *is in* $OP\mathcal{H}^{m_1 + m_2}(\Sigma)$.

Proof. Let $\Gamma_1 = \Gamma_2 = \Sigma^\#$ in (9.7). It is easy to verify that (9.7) is a clean diagram with excess d. Therefore by Theorem 9.5 the Schwartz kernel of $K_1 \circ K_2$ lies in $I^p(\Sigma^\#)$ where

$$p = m_1 + m_2 + (\dim X - d) - \frac{1}{2}(\dim X - d) = m_1 + m_2 + \frac{1}{2}(\dim X - d)\,.$$

Q.E.D.

All the results above can be sharpened by giving explicit formulas for the symbols involved. Unfortunately most of these formulas are not much simpler than the general formula given in §7. The composition formula in Theorem 9.6 *is* however, appreciably simpler if Σ is symplectic; that is, if the symplectic form $\sum dx_i \wedge d\zeta_i$ restricted to Σ is nondegenerate (thus defining a symplectic structure on Σ).

The normal space to Σ at each point can then be identified with the orthocomplement of the tangent space, and it too acquires a symplectic structure. Suppose that X has a metalinear structure. Then T^*X can be given a metaplectic structure, as we observed in §5. We will show that this induces a metaplectic structure on Σ. What amounts to the same thing, we will prove

PROPOSITION 9.7. *The normal bundle to Σ has a canonical metaplectic structure.*

Proof. Let $\Sigma^\#$ be, as above, the set

$$\{(x,\zeta,x,-\zeta),(x,\zeta)\in\Sigma\}\,.$$

$\Sigma^\#$ is diffeomorphic to Σ and is an isotropic submanifold of $T^*(X\times X)$. Let p be a point of $\Sigma^\#$. The metaplectic structure on $T_p(X\times X)$ gives us a "metalinear structure on $\Sigma^\#_p$ times a metaplectic structure on $(\Sigma^\#_p)^\perp/\Sigma^\#_p$." (See §5.) Since we have a canonical volume form on Σ_p the metalinear structure on $\Sigma^\#_p$ can be canonically trivialized; so we have a canonical metaplectic structure on $(\Sigma^\#_p)^\perp/\Sigma^\#_p$. Let N_p be the normal space to Σ at p. There is a canonical isomorphism

$$N_p \oplus N_p \xrightarrow{\cong} (\Sigma^\#_p)^\perp/\Sigma^\#_p \tag{9.9}$$

namely the inclusion map $N_p \oplus N_p \to T^*_p(X\times X)$ (whose image is in $(\Sigma^\#_p)^\perp$)

followed by projection. Moreover, (9.9) is a symplectic isomorphism provided we give $N_p \oplus N_p$ the symplectic structure of a direct sum. Let $\mathcal{B}$ be the set of symplectic bases of N_p and $\mathcal{B}^{\#}$ the set of symplectic bases of $(\Sigma_p^{\#})^{\perp}/\Sigma_p^{\#}$. Let $\mathfrak{M}^{\#}$ be the metaplectic double covering of $\mathcal{B}^{\#}$. Given a symplectic basis of N_p, we get a symplectic basis of $N_p \oplus N_p$ by taking two copies of it. Thus, there is an imbedding, $\mathcal{B} \to \mathcal{B}^{\#}$. Let $\mathfrak{M}$ be the subset of $\mathfrak{M}^{\#}$ lying above the image of $\mathcal{B}$. Then $\mathfrak{M} \to \mathcal{B}$ is the required metaplectic double covering of $\mathcal{B}$. Q.E.D.

Let H_p be the space of symplectic spinors associated with N_p and $H_p^{\#}$ the space of symplectic spinors associated with $(\Sigma_p^{\#})^{\perp}/\Sigma_p^{\#}$. As a corollary of Proposition 9.7, we get

$$H_p \hat{\otimes} H_p \xrightarrow{\cong} H_p^{\#} . \tag{9.10}$$

the tensor product being, of course, the Hilbert space tensor product. Using (9.10) we can identify $H_p^{\#}$ with the algebra of Hilbert-Schmidt operators on H_p. Then $(H_p^{\#})^{\infty}$, the space of C^{∞} vectors in $H_p^{\#}$, gets identified with a subalgebra which we will denote by SOP_p (an acronymn for "Schwartz operators"). The vector bundle $p \to SOP_p$ will be denoted by $SOP(\Sigma)$, the space of smooth sections of this bundle by $\mathcal{SOP}(\Sigma)$ and the sections which are homogeneous of degree m by $\mathcal{SOP}^m(\Sigma)$.

Let K be a Hermite operator of order m and let $\mathscr{k}$ be its Schwartz kernel. By (9.8) $\mathscr{k}$ belongs to the space $I^{m+(\dim X-d)/2}(\Sigma^{\#})$, so its symbol is a homogeneous section of $\mathrm{Spin}(\Sigma^{\#})$ of degree of homogeneity $m+(\dim X-d)/2$. At each point p of $\Sigma^{\#}$, this symbol is the product of a half-form on Σ_p and an element of $(H_p^{\#})^{\infty}$. The symplectic structure on Σ gives us, ready-at-hand, a non-vanishing half-form on Σ of degree of homogeneity $(\dim X-d)/2$; so, if we divide the symbol of $\mathscr{k}$ by this half-form we end up with an element of $\mathcal{SOP}^m(\Sigma)$. Thus we are provided with a canonical symbol map

$$OP\mathcal{H}^m(\Sigma) \to \mathcal{SOP}^m(\Sigma) . \tag{9.11}$$

THEOREM 9.8. *If* $K_1 \in OP\mathcal{H}^{m_1}$ *and* $K_2 \in OP\mathcal{H}^{m_2}$ *then* $\sigma(K_1)\sigma(K_2) = \sigma(K_1K_2)$.

Proof. Let k_1 and k_2 be the corresponding Schwartz kernels, and let k be the Schwartz kernel of K_1K_2. We can write

$$k = \pi_*\Delta^* k_1 \boxtimes k_2$$

where $\Delta : X \times X \times X \to X \times X \times X \times X$ is the diagonal map, $(x,y,z) \to (x,y,y,z)$; and $\pi : X \times X \times X \to X \times X$ is the projection map, $\pi(x,y,z) = (x,z)$. By (9.2) the symbol of k is

$$\pi_*\Delta^*\sigma(k_1) \boxtimes \sigma(k_2)\,. \tag{9.12}$$

Let p be a point of Σ and $a_i = \sigma(K_i)_p$, $i = 1,2$. Then a_i is in $(SOP)_p$. Modulo the identifications above a_i is a Hilbert-Schmidt operator on H_p whose kernel is $\sigma(k_i)_p$. The formula (9.12) is just the usual formula for the kernel of the composition of a_1 and a_2. Thus the product symbol is the composition of the symbols, as asserted.

We conclude this section with some remarks about the parity properties of the distributions in $I^m(X,\Sigma)$.

Let $u \in I^m(X,\Sigma)$ be defined by an oscillatory integral of the form (3.4) with an amplitude of the form (3.2) where the degrees of homogeneity, m_i, of the $a_i(\tau,\eta)$'s are all either integers or half-integers. We will say that u is *even* if

$$a_i(\tau,\eta) = (-1)^{2m_i - N} a_i(\tau,-\eta) \tag{9.13$_{even}$}$$

for all i, and *odd* if

$$a_i(\tau,\eta) = (-1)^{2m_i - N+1} a_i(\tau,-\eta) \tag{9.13$_{odd}$}$$

for all i. One can show that this definition is independent of the way in which u is written as an oscillatory integral. (For instance it is perfunctory to check that the change of coordinates formulas in §9 of [15] behave correctly with respect to parity.) By inspecting the parities of the amplitudes in (8.9) and (8.10) one easily establishes the following sharpened form of Theorem 7.5.

THEOREM 9.9. *Suppose* Γ, Σ *and* K *satisfy the hypotheses of Theorem 7.5. Then if* $(\dim Y - e)/2$ *is an integer,* $K: I^s(\Sigma) \to I^{r+s-1/2(\dim y-e)}(\Gamma \circ \Sigma)$ *preserves parity and if it is a half-integer,* K *reverses parity.*

In particular, if Q is a pseudodifferential operator of order k on X, with k an integer, then $Q: I^r(\Sigma) \to I^{r+k}(\Sigma)$ preserves parity.

We will next describe how this notion of parity reflects itself in the symbol calculus.

The metaplectic representation

$$M_p(n) \to U(H)$$

is not irreducible; however it is well known that H breaks up into two irreducible subspaces

$$H = H_{\text{even}} \oplus H_{\text{odd}} \tag{9.14}$$

each of which is irreducible. (These are the symplectic analogues of the odd and even spinors in the Clifford algebra.) Corresponding to (9.14) we get a direct sum of vector bundles

$$\operatorname{Spin}(\Sigma) = \operatorname{Spin}(\Sigma)_{\text{even}} \oplus \operatorname{Spin}(\Sigma)_{\text{odd}} .$$

We will say that a homogeneous section of $\operatorname{Spin}(\Sigma)_{\text{even}}$ is *even* (or *odd*) if its order of homogeneity is integral (or half-integral). Similarly we will say that a homogeneous section of $\operatorname{Spin}(\Sigma)_{\text{odd}}$ is odd (or even) if its order of homogeneity is integral (or half-integral). With these conventions the symbol map

$$I^r(X, \Sigma) \to S^r(\Sigma)$$

preserves parity, as one easily sees just by inspection.

One trivial but useful observation is the following. Suppose $\Sigma \subset T^*X$ is actually a Lagrangian manifold, i.e. *maximal* isotropic. Then in the symbolic expansion (3.2) only τ-variables, i.e. only "non-damped" variables occur, so the only possible parity combinations in (9.13) are $2m_i - N \equiv 0 \mod 2$ for even distributions and $2m_i - N \equiv 1 \mod 2$ for odd distributions. In other words the filtration of I_{even} by degree goes down by *unit* stops at a time rather than half-unit steps, and the same is true of the filtration of I_{odd}. Therefore I_{even} and I_{odd} are the usual spaces of Fourier integral distributions of classical type. For $u \in I_{even}$ the total symbol of u is a sum of homogeneous terms of integral degree and for $u \in I_{odd}$ of half-integral degree.

This remark has one important consequence for us. Let X and Y be smooth manifolds with $\dim X < \dim Y$, and let $\Phi : T^*X - 0 \to T^*Y - 0$ be a homogeneous symplectic imbedding, i.e. an imbedding such that the pull-back by Φ of the canonical symplectic form on $T^*X - 0$ is the canonical symplectic form on $T^*X - 0$. Let

$$\Sigma_\Phi = \{(x, \xi, y, \eta), (y, -\eta) = \Phi(x, \xi)\} .$$

Let $\mathcal{T}$ be an element of $I^r(X \times Y, \Sigma_\Phi)$ and let F be the operator having $\mathcal{T}$ as its Schwartz kernel. Using Theorem 9.9 and arguments similar to those in the proof of Theorem 9.5 one can prove that the Schwartz kernel of F^*F belongs to $I^{2r+n/2}_{even}(X \times X, \Delta)$ where $n = \dim Y$ and $\Delta = \{(x,\xi,x,-\xi), (x,\xi) \in T^*X - 0\}$. In other words F^*F is a classical pseudodifferential operator of order $2r + (\dim Y - \dim X)/2$. Its total symbol is a sum of homogeneous terms of order of homogeneity, $n/2 + s$, s integer. More generally if Q is a pseudodifferential operator of classical type on Y then F^*QF is a pseudodifferential operator of classical type on X whose total symbol involves either integer orders of homogeneity or integer orders of homogeneity shifted by one-half. By a symbolic computation, based on the second part of Theorem 7.5, one easily verifies that the symbol of F^*QF is the symbol of F^*F times $\Phi^*\sigma(Q)$.

Let us apply these remarks to the operator, R, of §2. From the expression (2.7) one sees that the Schwartz kernel of R belongs to the space $I^{-n/2}(\mathbf{R}^q \times \mathbf{R}^q, \Sigma_\Phi)$ where Φ is the canonical transformation $(t,\tau) \in T^*\mathbf{R}^q - 0 \to (t,\tau,0,0)$ imbedding $T^*\mathbf{R}^n - 0$ as the submanifold, $y = \eta = 0$, of $T^*\mathbf{R}^n - 0$. Hence if Q is a classical pseudodifferential operator of order k on $\mathbf{R}^n$, R^*QR is a classical pseudodifferential operator of order k on $\mathbf{R}^q$, its leading symbol being the restriction of the leading symbol of Q to the set, $y = \eta = 0$. (See Proposition 2.3.)

§10. THE TRANSPORT EQUATION

One important special case of Theorem 7.5 is the following.

THEOREM 10.1. *Let* P *be a pseudodifferential operator of order* m, *operating on half-forms. Then* P *maps* $I^k(X,\Sigma)$ *into* $I^{k+m}(X,\Sigma)$. *Moreover, if* p_m *is the symbol of* P, *then, for* $u \in I^k(X,\Sigma)$, $\sigma(Pu) = p_m\sigma(u)$.

A corollary of this theorem is that if $p_m = 0$ on Σ then $Pu \in I^{k+m-1/2}(X,\Sigma)$, and we can ask what the symbol of Pu is now. In order to state the answer we need to go back and look a little more carefully at the metaplectic representation. Let $\mathcal{S}$ be the Schwartz space on which the n-dimensional metaplectic group acts. $\mathcal{S}$ is also a representation space for the Heisenberg algebra, n. (See §4.) Moreover if $\tau : Mp(n) \to \mathrm{Hom}(\mathcal{S})$ is the metaplectic representation and $d\rho : n \to \mathrm{Hom}(\mathcal{S})$ the Heisenberg representation, we have

$$d\rho(Av) = \tau(A)d\rho(v)\tau(A)^{-1} . \tag{10.1}$$

(Compare with (4.4).)

Let Z be a metaplectic manifold and Σ an isotropic submanifold. Let E be the symplectic normal bundle of Σ. For each $x \in \Sigma$ we can make $E_x \oplus R$ into a Heisenberg algebra in an obvious way. Namely if Ω^{E_x} is the symplectic two form on E_x we set

$$[u,v] = \Omega^{E_x}(u,v)]$$
$$[1,\text{anything}] = 0$$

where 1 is the unit vector in R and $u,v \in E_x$. Now choose a metaplectic basis of E_x and a half form, μ, in the tangent space to Σ at x.

This gives us an identification of $E_x \oplus R$ with n and an identification of $\mathrm{Spin}(\Sigma)_x$ with $\mathcal{S}$. Given $v \in E_x$ and $\sigma \in \mathrm{Spin}(\Sigma)_x$ we define $v\sigma$ to be $(d\rho)(v)\sigma \in \mathrm{Spin}(\Sigma)_x$. By (10.1) the definition of $v\sigma$ is independent of our choice of metaplectic basis. It is also clearly independent of our choice of μ. Therefore, *there is a canonical representation of the Heisenberg algebra* $E_x \oplus R$ *on* $\mathrm{Spin}(\Sigma)_x$ *given by* $(v, \sigma) \to v\sigma$.

Let p be a function on Z which vanishes identically on Σ. Let Ξ_p be its associated Hamiltonian vector field. If $x \in \Sigma$, let Σ_x be the tangent space to Σ at x. If $v \in \Sigma_x$ then $\Omega(v, \Xi_p) = \langle v, dp \rangle = 0$; so $\Xi_p(x)$ is in $\Sigma_x^\perp$ at x. This means that Ξ_p defines a section v_p of the symplectic normal bundle E. Moreover, given a section σ of $\mathrm{Spin}(\Sigma)$ we get a new section $v_p\sigma$ by the procedure outlined above.

Now let Σ be a closed homogeneous isotropic submanifold of T^*X-0. We refer to §10 of [15] for the proof of the following.

THEOREM 10.2. *Let* P *be a pseudodifferential operator of order* m *with symbol* $p = p_m$. *Suppose* p *vanishes on* Σ. *Let* $u \in I^k(X, \Sigma)$. *Then* $Pu \in I^{m+k-1/2}(X, \Sigma)$ *and*

$$\sigma(Pu) = (1/\sqrt{-1})v_p\sigma(u) . \tag{10.2}$$

REMARK. The operator v_p occurs in Hörmander [23] as the test operator for testing subelliptic estimates of order $-1/2$. If p has simple characteristics it is not hard to derive Hörmander's results on subellipticity of order $-1/2$ from our results by purely formal (i.e. symbolic) arguments.

Suppose the Hamiltonian vector field Ξ_p is tangent to Σ everywhere. This means that v_p is identically zero and, therefore, $Pu \in I^{m+k-1}(\Sigma)$. What is its symbol? To answer this question observe that the Hamiltonian flow associated with p maps Σ onto itself and so maps $\mathrm{Spin}(\Sigma)$ onto itself. Therefore one can differentiate sections of $\mathrm{Spin}(\Sigma)$ with respect to Ξ_p. We will denote the derivative of such a section, say σ, by $\Xi_p\sigma$. It is easy to see that $\Xi_p\sigma$ is in $S^{m+k-1}(\Sigma)$ when σ is in $S^k(\Sigma)$.

For the following we again refer to §10 of [15].

THEOREM 10.3. *If* Ξ_p *is tangent to* Σ *then*

$$\sigma(Pu) = (1/\sqrt{-1})\Xi_p\sigma(u) + p_{sub}\sigma(u) . \tag{10.3}$$

p_{sub} *being the subprincipal symbol of* u.

Note that when Σ is Lagrangian, (10.3) is just the transport equation of Duistermaat-Hörmander [13].

§11. SYMBOLIC PROPERTIES OF TOEPLITZ OPERATORS

In this section we will show that Toeplitz operators of the generalized kind introduced in §2 are Hermite operators. We will also discuss the notion of principal and subprincipal symbol for Toeplitz operators (and show that the latter is not entirely a well-defined concept). As in §2 let X be a compact manifold and Σ a closed, homogeneous symplectic submanifold of $T^*X - 0$. We will assume that X has a metalinear structure, so that the notion of half-form makes sense. Given half-forms μ and ν, the product $\mu\bar{\nu}$ is a density and its integral over X

$$\int \mu\bar{\nu} \tag{11.1}$$

is well defined since X is compact. We will denote by $L^2 = L^2(X)$ the space of L^2-integrable half-forms equipped with the Hilbert space inner product given by (11.1).

Let there be given a Toeplitz structure on Σ defined by an operator $\pi_\Sigma : L^2 \to L^2$ which satisfies $\pi_\Sigma = \pi_\Sigma^2 = \pi_\Sigma^*$ and is described microlocally by condition III of definition 2.10.

THEOREM 11.1. *π_Σ belongs to the ring* $OP\mathcal{H}^o(\Sigma)$.

Proof. By definition 2.10 it is enough to prove this assertion for the operator π of Proposition 2.1. By (2.2) the Schwartz kernel of π is

$$\int e^{i(y\eta+y'\nu'+(t-t')\tau)}(\tau/\pi)^{-p/2}\, e^{-(1/2|\tau|)(\eta^2+(\eta')^2)}\, d\tau\, d\eta\, d\eta' . \tag{11.2}$$

One sees by inspection that this is the Schwartz kernel of a Hermite operator associated with the symplectic submanifold, $y = \eta = 0$, in T^*R^n. Q.E.D.

To describe the symbol of the operator π_Σ we first recall some results from §4. Let V be a symplectic vector space, let $\mathcal{S}$ be the space of C^∞ vectors in the representation space for the metaplectic representation of Mp(V) and let Λ be a positive definite Lagrangian subspace of $V \otimes \mathbf{C}$. We showed in §4 that Λ determines, uniquely up to a constant multiple of modulus one, a "vacuum state" e_Λ in $\mathcal{S}$. In turn, given a vacuum state, e, in $\mathcal{S}$ there is a unique Λ such that $e = e_\Lambda$. In the paragraph above let z be a point on Σ and let V be the normal space to Σ at z. We recall from §9 that if T is in $OP\mathcal{H}^m(\Sigma)$ then the symbol of T at z is itself an operator, going from $\mathcal{S}$ to $\mathcal{S}$. Now given a positive definite Lagrangian subspace, Λ_z, of V there is a canonical such operator associated with it, namely "projection onto the one-dimensional space spanned by e_Λ." Let us denote this operator by π_Λ.

THEOREM 11.2. *There is canonically associated with π_Σ a homogeneous positive definite Lagrangian subbundle*

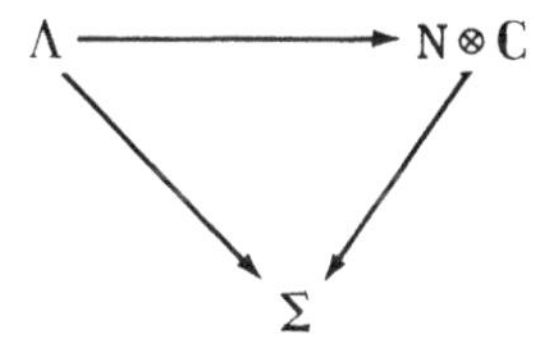

of the complexified normal bundle, $N \otimes \mathbf{C}$, *of* Σ *such that* $\sigma(\pi_\Sigma) = \pi_\Lambda$ *at each point* $z \in \Sigma$.

Proof. It is enough to prove this theorem for the operator π of (2.2). Fix a point $z \in \Sigma_0$. The symbol of π at z is an operator $\pi' : \mathcal{S} \to \mathcal{S}$ as we pointed out in the paragraph above. Since π satisfies $\pi^2 = \pi = \pi^*$, so does π' because of Theorem 9.8. By definition, $D_i \pi = 0$, D_i being the operator (2.1). Let Ξ_i be the Hamiltonian vector field on $T^*R^n - 0$ having the symbol of D_i as its associated Hamiltonian function. At each point, z, of Σ the Ξ_i's span a positive definite Lagrangian subspace, Λ_z, of the normal space to Σ at z. Applying Proposition 10.2 to the operator equation, $D_i \pi = 0$, at z we see that the range of π' at z is one-dimensional and is spanned by the vacuum state, e_Λ. Q.E.D.

Let Q be a pseudodifferential operator mapping the space of half-forms on X into itself. Combining Theorem 11.2 with results from §9 we easily deduce

THEOREM 11.3. *The operator* $\pi_\Sigma Q \pi_\Sigma$ *is in* $\mathrm{OP}\mathcal{H}^m(\Sigma)$ *and its symbol is* $q\pi_\Lambda$, q *being the symbol of* Q.

REMARK. In §2 we defined the symbol of $\pi_\Sigma Q \pi_\Sigma$, qua Toeplitz operator, to be the symbol of Q restricted to Σ. We see above that modulo the identification $q \to q\pi_\Lambda$ this definition accords with the definition of the symbol of $\pi_\Sigma Q \pi_\Sigma$ as a Hermite operator. It also shows again that the symbol of $\pi_\Sigma Q \pi_\Sigma$ is well defined. (See Proposition 2.3 and the end of §9.)

We will next discuss the notion of "subprincipal symbol" for Toeplitz operators. At this point it will be convenient to introduce the ring of all pseudodifferential operators on X which commute with π_Σ. We will denote this ring by $\mathcal{R}$. This ring contains as an ideal, $\mathcal{I}$, the set of all $Q \in \mathcal{R}$ such that $Q\pi_\Sigma = 0$. In §2 we showed that every Toeplitz operator can be written in the form $\pi_\Sigma Q \pi_\Sigma$ with $Q \in \mathcal{R}$; so the ring of Toeplitz operators is just the quotient ring $\mathcal{R}/\mathcal{I}$.

PROPOSITION 11.4. *Let* Q *be in* $\mathcal{R}$ *and let* q *be its symbol. Let* H_q *be the Hamiltonian vector field associated with* q. *Then, at each point of* Σ, H_q *is tangent to* Σ.

Proof. Let v be the projection of $H_q(z)$ in the normal space to Σ at z. Applying Theorem 10.2 to the equation, $Q\pi_\Sigma - \pi_\Sigma Q = 0$ we get the following symbolic equation at z:

$$ve_\Lambda \otimes (e_\Lambda, -) + e_\Lambda \otimes (\bar{v} e_\Lambda, -) = 0, \tag{11.3}$$

from which we deduce that ve_Λ is a constant multiple of e_Λ. However, e_Λ is an "even" element of the metaplectic representation space and ve_Λ an "odd" element, so $ve_\Lambda = 0$. By Proposition 4.2, $v \in \Lambda$. From (11.3) we also deduce that $\bar{v}e_\Lambda = 0$; so $\bar{v} \in \Lambda$. Since $\Lambda \cap \bar{\Lambda} = 0$, $v = 0$.

Q.E.D.

REMARK. One consequence of this theorem is that the Hamiltonian vector field, H_q, restricted to Σ is identical with the Hamiltonian vector field associated with the function $q|\Sigma$ (the latter being defined by means of the intrinsic symplectic structure on Σ).

Suppose now that the symbol, q, of Q is real-valued. Then H_q is a real vector field; so it is the infinitesimal generator of a one-parameter group, $\exp tH_q$, of symplectic diffeomorphisms. By Proposition 11.4 $\exp tH_q$ maps Σ into Σ; so if N is the normal bundle of Σ, $\exp tH_q$ induces a mapping of N into N.

PROPOSITION 11.5. $\exp tH_q$ *maps* Λ *into* Λ.

Proof. Let $e_\Lambda(z)$ be a "vacuum state" associated with the Lagrangian subspace Λ_z of N_z. Given $z_o \in \Sigma$ we can find a neighborhood, U, of z_o in Σ such that on U, $e_\Lambda(z)$ depends smoothly on z. Applying Theorem 10.3 to the operator equation, $Q\pi_\Sigma - \pi_\Sigma Q = 0$ we obtain the following symbolic equation:

$$H_q e_\Lambda \otimes (e_\Lambda, -) - e_\Lambda \otimes (H_q e_\Lambda, -) = 0 . \tag{11.4}$$

From (11.4) we deduce that $H_q e_\Lambda = f e_\Lambda$ on U, for some $f \in C^\infty(U)$. Let $z = (\exp tH_q)(z_o)$. The map, $d(\exp tH_q): N_{z_o} \to N_z$ induces a map from the space of symplectic spinors at z_o into the space of symplectic spinors at z. Integrating the equation, $H_q e_\Lambda = f e_\Lambda$, along the trajectory going from z_o to z we see that this map maps $e_\Lambda(z_o)$ into a multiple of $e_\Lambda(z)$. From Proposition 4.3 we conclude that $d(\exp tH_q)_{z_o}$ maps Λ_{z_o} into Λ_z.

Q.E.D.

As above let V be a fixed symplectic vector space, $\mathcal{S}$ the space of C^∞ vectors in the representation space of the metaplectic representation of $Mp(V)$ and Λ a positive definite Lagrangian subspace of $V \otimes \mathbb{C}$. Let U_Λ be the subgroup of $Mp(V)$ leaving Λ fixed and let u_Λ be its Lie algebra. Finally let $e_\Lambda \in \mathcal{S}$ be a vacuum state associated with Λ. Given $A \in U_\Lambda$, $Ae_\Lambda = c(A)e_\Lambda$ for some complex number $c(A)$ depending only on A. In §4 we showed that c is a unitary character of the group U_Λ. We will denote by $dc : u_\Lambda \to \sqrt{-1}\,\mathbb{R}$ the infinitesimal character associated with c.

Now let Q be an element of $\mathcal{J}$ and let q be the leading symbol of Q. Let z be a point of Σ and let V be the normal space to Σ at z. By Theorem 11.3 $q = 0$ on Σ, and by Proposition 11.4 and the remark following it $H_q = 0$ on Σ; so q vanishes on Σ together with all its first derivatives. Therefore the Hessian of q at z is intrinsically defined as a quadratic form on V. As we saw in §4, to each such quadratic form is canonically associated a linear symplectic mapping $A_q : V \to V$. With the notation above we will prove

PROPOSITION 11.6. A_q *belongs to* u_Λ. *Moreover*

$$(1/\sqrt{-1})\,dc(A_q) + \sigma_{sub}(Q)(z) = 0\,, \tag{11.5}$$

$\sigma_{sub}(Q)$ *being the value of the subprincipal symbol of* Q *at* z.

Proof. Since $Q\pi_\Sigma = 0$ we get from Theorem 10.3

$$(1/\sqrt{-1})\,A_q e_\Lambda \otimes (e_\Lambda, -) + \sigma_{sub}(Q)\, e_\Lambda \otimes (e_\Lambda, -) = 0 \tag{11.6}$$

at z. This equation first of all tells us that $A_q e_\Lambda$ is a multiple of e_Λ; hence that $A_q \in u_\Lambda$. Since $A_q \in u_\Lambda$, $A_q e_\Lambda = dc(A_q)e_\Lambda$, and (11.5) follows from (11.6). Q.E.D.

Given $z \in \Sigma$ let us denote by SV_z the set of all vacuum states associated with the Lagrangian subspace Λ_z. The assignment $z \to SV_z$

defines a circle bundle

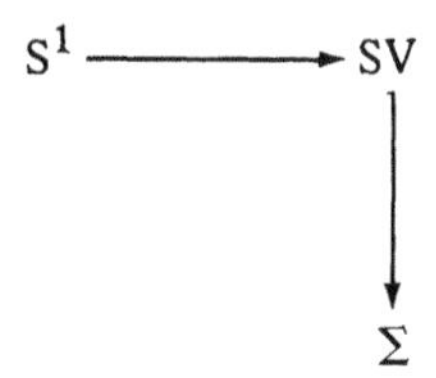

For the moment let us suppose that this bundle is trivial. Let $e_\Lambda : \Sigma \to SV$ be a section of SV which is homogeneous of order zero. With e_Λ fixed we will define the notion of *subprincipal symbol* for Toeplitz operators as follows. Given $T \in \mathcal{R}/\mathcal{I}$ let Q be a representative of T in $\mathcal{R}$ and let q be its symbol. By (11.4) $H_q e_\Lambda = c(Q) e_\Lambda$, $c(Q)$ being a smooth function on Σ. We will define the subprincipal symbol of T to be the expression

$$\sigma_{sub}(T) = (1/\sqrt{-1})c(Q) + \sigma_{sub}(Q) \tag{11.6}$$

$\sigma_{sub}(Q)$ being the subprincipal symbol of Q. To show that (11.6) depends only on T, not on Q, let Q_1 be another representative of T in $\mathcal{R}$, and let $Q_2 = Q - Q_1$. Then $Q_2 \in \mathcal{I}$ and

$$(1/\sqrt{-1})c(Q_2) + \sigma_{sup}(Q_2)$$

is equal to the expression (11.5) at $z \in \Sigma$. Since this expression vanishes, the right-hand side of (11.6) is unambiguously defined.

To define the subprincipal symbol it is not quite necessary to assume that the bundle SV is trivial. It is sufficient to assume the following:

(*) There exists a covering of Σ by conic open sets $\{U\}$ and for each U a homogeneous section $e_U : U \to SV$ such that the transition functions, $f_{UV} : e_V = f_{UV} e_U$ are constants.

Indeed if this is the case then $H_q e_U = C(Q) e_U \Longrightarrow H_q e_V = c(Q) e_V$ on the overlap $U \cap V$; so $c(Q)$ is well defined. We will show that (*) is satisfied providing Λ satisfies the following topological condition.

PROPOSITION 11.7. *Let* $c = c_1(\Lambda) \in H^2(\Sigma, Z)$ *be the first Chern class of the complex vector bundle* $\Lambda \to \Sigma$. *Let* c' *be the image of* c *in* $H^2(\Sigma, R)$. *Then there exists a family of local sections of* SV *satisfying* (*) *if and only if* $c' = 0$.

Proof. The first step in the proof will be to give an alternative description of the circle bundle, SV. As we pointed out in §4 the vector bundle $\Lambda \to \Sigma$ has a canonical inner product. Let

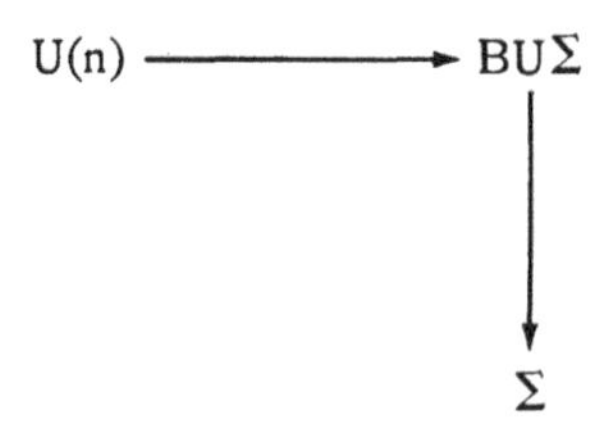

be the principal U(n) bundle whose fiber above z is the set of all orthonormal bases of Λ_z. Let $Bp\Sigma$ be the symplectic basis bundle associated with the symplectic vector bundle $N \to \Sigma$. Let $Mp\Sigma$ be the metaplectic double covering of $Bp\Sigma$. (See Proposition 9.7.) There is a canonical imbedding $BU\Sigma \to Bp\Sigma$ which assigns to the orthogonal basis $e_1, \cdots, e_n$ of Λ_z the symplectic basis $(\text{Re } e_1, \cdots, \text{Re } e_n, \text{Im } e_1, \cdots, \text{Im } e_n)$; so we can think of $BU\Sigma$ as sitting inside of $Bp\Sigma$. If $MU\Sigma$ is the double covering of $BU\Sigma$ in $Mp\Sigma$ and MU(n) the double covering of U(n) in Mp(n) then we have a principal fibration

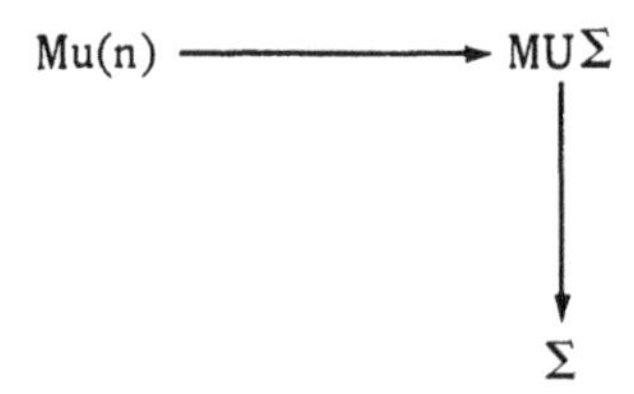

By Proposition 4.3 the character, $\det : U(n) \to S^1$ has a canonical square root, $\sqrt{\det}$ on MU(n). By means of $\sqrt{\det}$, we can associate with $MU\Sigma$ a circle bundle, SM, over Σ.

PROPOSITION 11.8. *The circle bundles* SV *and* SM *are canonically isomorphic.*

Proof. This is a corollary of Proposition 4.3. Q.E.D.

Coming back to the proof of Proposition 11.7, let SB be the circle bundle associated to $BU\Sigma$ by means of the character $\det : U(n) \to S^1$. The bundle SM admits a family of local sections satisfying (*) if and only if the same is true of the bundle SB; so by Proposition 11.8 it is enough to prove Proposition 11.7 with SV replaced by SB. The Chern class of SB is the Chern class of the complex "volume form" bundle $\wedge^n \Lambda$ and this in turn is just the first Chern class of Λ itself. Its image, c', in $H^2(\Sigma, \mathbf{R})$ can be computed as follows: (see [36]). Let α be a connection form on SB and let $R(\alpha)$ be its curvature form. Then c' is equal to the De Rham class determined by $R(\alpha)$. If $c' = 0$ there exists a one-form α_o on Σ such that $d\alpha_o + R(\alpha) = 0$. Replacing α by $\alpha + \alpha_o$ we can arrange that $R(\alpha) = 0$. Now choose the e_U's to be sections of SB which are invariant under parallel transport. These sections will satisfy all the hypotheses of (*) except the homogeneity. To make them homogeneous let Σ_o be a codimension one submanifold of Σ such that the homothety map $\Sigma_o \times \mathbf{R}^+ \to \Sigma$ is a diffeomorphism. Restrict the e_U's to Σ_o then extend them again by homogeneity. In the other direction given a family $\{e_U\}$ satisfying (*) there is a unique connection, α, on SB such that $R(\alpha) = 0$ and such that the e_U's are flat sections of SB. Q.E.D.

Proposition 11.7 shows that the class of those manifolds for which the subprincipal symbol can be defined is rather restrictive. We also need to emphasize that the definition of the subprincipal symbol depends very much on the choice of the family of local sections $\{e_U\}$ in (*).

We will next show that the subprincipal symbol satisfies the usual composition law.

PROPOSITION 11.9. *If* T_1 *and* T_2 *are in* $\mathcal{R}/\mathcal{I}$ *then*

(11.7) $\sigma_{\mathrm{sub}}(T_1 T_2) = \sigma(T_1)\sigma_{\mathrm{sub}}(T_2) + \sigma(T_2)\sigma_{\mathrm{sub}}(T_1) + (1/2\sqrt{-1})\{\sigma(T_1), \sigma(T_2)\}$.

Proof. Let Q_i, $i = 1,2$, be a representative of T_i in $\mathcal{R}$ and let q_i be its symbol. Then $H_{q_1 q_2} = q_1 H_{q_2} + q_2 H_{q_1}$; so $c(Q_1 Q_2) = \sigma(Q_1)c(Q_2) + \sigma(Q_2)c(Q_1)$. This, together with the usual composition law for $\sigma_{\mathrm{sub}}(Q_1 Q_2)$ yields (11.7). Q.E.D.

Even when the subprincipal symbol cannot be defined we can still define a certain vestigial piece of it. First, assuming σ_{sub} can be defined on $\mathcal{R}/\mathcal{S}$ in the manner outlined above, we will prove

PROPOSITION 11.10. *Let* T *be a Toeplitz operator and let* $\sigma(T)$ *and* $\sigma_{\mathrm{sub}}(T)$ *be the principal and subprincipal symbols of* T *respectively. Assume* $\sigma(T)$ *is real-valued and let* Ξ *be the Hamiltonian vector field associated with it. Let* γ *be a closed integral curve of* Ξ. *Then the expression*

$$\exp \sqrt{-1}\left(\int_\gamma \sigma_{\mathrm{sub}}(T)\right) \tag{11.8}$$

is an invariant of T *alone, not depending on the choice above of the family of local sections.*

(Note that the expression (11.8) is precisely the contribution of the subprincipal symbol of T to the trace formula (1.5).)

Proof. Choose a representative, Q, for T in $\mathcal{R}$ and let q be the symbol of Q. By the remark after Proposition 11.4, $\Xi = H_q|\Sigma$. If e_Λ and e'_Λ are non-vanishing global sections of SV then $e'_\Lambda = he_\Lambda$ for some smooth function $h : \Sigma \to S^1$. If $H_q e_\Lambda = c(Q)e_\Lambda$ then $H_q e'_\Lambda = c(Q)he_\Lambda + Hhe_\Lambda$; so

$$c'(Q) = c(Q) + H_q \arg h$$

where $H_q e'_\Lambda = c'(Q) e'_\Lambda$. Therefore, with respect to the trivialization, e'_Λ, the left-hand side of 11.6 becomes $\sigma_{sub}(T) + H_q \arg h$. The integral of the second term over γ is just 2π times the winding number of the map $h : \gamma \to S^1$; so the exponential of the product of this integral with $\sqrt{-1}$ is zero. Q.E.D.

This same proof shows that the quantity (11.8) is well defined even when there is neither a global section, e_Λ, of SV nor system of local sections $\{e_U\}$ satisfying (*). We will continue to use the suggestive notation on the right-hand side of (11.8) for this quantity even though this notation is a little misleading.

§12. THE TRACE FORMULA

Let X be a compact, metalinear manifold and Σ a homogeneous symplectic submanifold of $T^*X - 0$, and let $\pi : L^2(X) \to L^2(X)$ define a Toeplitz structure on Σ. Let T be a self-adjoint first order Toeplitz operator such that $\sigma(T) > 0$ everywhere on Σ. By Proposition 2.14, the spectrum of T is discrete, bounded below and has only $+\infty$ as an accumulation point. Let $N(\lambda)$ be the number of eigenvalues of T which are less than λ.

PROPOSITION 12.1. *There exists a constant* $m > 0$ *such that* $N(\lambda) \leq \lambda^m$ *for* λ *large.*

Proof. We will first of all prove:

LEMMA 12.2. *There exists a self-adjoint first order pseudodifferential operator* $Q : L^2(X) \to L^2(X)$ *such that* $[Q, \pi] = 0$, $\sigma(Q) > 0$ *everywhere and* $T = \pi Q \pi$.

Proof. By Proposition 2.12 there exists a self-adjoint pseudodifferential operator Q_o such that $[\pi, Q_o] = 0$ and $T = \pi Q_o \pi$. Since $\sigma(Q_o) = \sigma(T)$ on Σ there exists a conic neighborhood $\mathcal{C}_1$ of Σ such that $\sigma(Q_o) > 0$ on $\overline{\mathcal{C}}_1$. Let $\mathcal{C}_o$ be a conic neighborhood of Σ such that $\mathcal{C}_2 \subset\subset \mathcal{C}_1$ and let R be a pseudodifferential operator of order zero such that $0 \leq \sigma(R) \leq 1$, R is smoothing outside $\mathcal{C}_1$, and $I - R$ is smoothing on $\mathcal{C}_2$. Let Q_1 be any pseudodifferential operator of order one whose symbol is everywhere positive and let $Q_2 = RQ_o + (I-R)Q_1$. Then $\sigma(Q_2) > 0$ everywhere and both $[Q_2, \pi]$ and $T - \pi Q_2 \pi$ are smoothing. Let $Q_3 = T + (1-\pi)Q_2(1-\pi)$. Q_3 differs from Q_2 by a smoothing operator, commutes with π and satisfies $\pi Q_3 \pi = T$. Finally let $Q = (Q_3 + Q_3^*)/2$. Q.E.D.

With Q as in Lemma 12.2, let $N_Q(\lambda)$ be the number of eigenvalues of Q less than λ. By standard theorems on elliptic pseudodifferential operators (e.g. [24]) there exists an $m > 0$ such that $N_Q(\lambda) \leq \lambda^m$ for λ large. Since $N(\lambda) \leq N_Q(\lambda)$ this proves Proposition 12.1. Q.E.D.

Let $\lambda_0 \leq \lambda_1 \leq \lambda_2 \leq \cdots$ be the spectrum of T. Consider the partial sum

$$e_\lambda(t) = \sum_{\lambda_i \leq \lambda} e^{\sqrt{-1}\,\lambda_i t}\,.$$

PROPOSITION 12.3. *As* $\lambda \to \infty$ $e_\lambda(t)$ *converges to a limit* $e(t)$, $e(t)$ *being a tempered distribution on* $\mathbf{R}$ *(and convergence being in the tempered distribution topology).*

Proof. The measure $f_\lambda(s) = \sum_{\lambda_i \leq \lambda} \delta(s-\lambda_i)$ converges to a tempered distribution by Proposition 12.1, hence so does $e_\lambda(t) = \hat{f}_\lambda(t)$.

Let Ξ be the Hamiltonian vector field on Σ associated with the symbol of T.

PROPOSITION 12.4. *The singular support of the distribution* $e(t)$ *is contained in the set of periods of the periodic orbits of* Ξ.

Proof. For simplicity assume that zero is not among the eigenvalues of T. Let Q be as in Lemma 12.2. We can assume that 0 is not among the eigenvalues of Q; so Q is invertible. For N sufficiently large Q^{-N} is an operator of trace class, hence so also is $Q^{-N}(\exp \sqrt{-1}\, tQ)\pi$. On the one hand the trace of this operator is

$$e_N(t) = \sum (\lambda_i)^{-N} e^{\sqrt{-1}\,\lambda_i t}\,. \tag{12.1}$$

On the other hand if $U_N = U_N(x,y,t)\sqrt{dx}\,\sqrt{dy}$ is the Schwartz kernel of this operator, then its trace is

$$\int U_N(x, x, t)\,dx\,. \tag{12.2}$$

Letting $\Delta : X \times \mathbf{R} \to X \times X \times \mathbf{R}$ be the diagonal map: $(x, r) \to (x, x, r)$ and $\pi : X \times \mathbf{R} \to \mathbf{R}$ the map: $(x, r) \to r$, then (12.2) is equal to $\pi_* \Delta^* U_N$. By [13], the wave front set of U_N is contained in the set of all $(x, \xi, y, \eta, t, \tau)$ such that $(x, \xi) \in \Sigma$ and (x, ξ) and $(y, -\eta)$ are joined by a bicharacteristic of Q of length t. The operator $\pi_* \Delta^*$ is a Fourier integral operator associated with the canonical relation $\{(t, \tau, x, \xi, x, \xi, t, \tau), (t, \tau) \in T^*\mathbf{R} - 0, (x, \xi) \in T^*X - 0\}$, so, by [25], the wave front set of $\pi_* \Delta^* U_N$ is contained in the set of all (t, τ) such that there exists an (x, ξ) which is joined to itself by a bicharacteristic of length t. Finally $e(t) = ((1/\sqrt{-1})(d/dt))^N e_N(t)$; so $e(t)$ and $e_N(t)$ have the same wave front sets. Q.E.D.

REMARK. Let U be the Schwartz kernel of $(\exp \sqrt{-1}\, tQ)\pi$. Though U is not of trace class $\pi_* \Delta^* U$ is well defined in the sense of distributions by [25], §2, π_* and Δ^* being the usual pull-back and push-forward operations on distributions. Since

$$((1/\sqrt{-1})(d/dt))^N Q^{-N} (\exp \sqrt{-1}\, tQ)\pi = (\exp \sqrt{-1}\, tQ)\pi$$

the argument above shows that

$$e(t) = \pi_* \Delta^* U\,. \tag{12.3}$$

We will henceforth use (12.3) as the *definition* of $e(t)$.

If the periodic integral curves of Ξ form "nice" submanifolds of Σ one can get sharper results on the nature of the singularities of $e(t)$. To describe these results we will need some symplectic preliminaries. First let M be a manifold and $\Phi : M \to M$ a diffeomorphism.

DEFINITION 12.5. A submanifold, Z, of fixed points of Φ will be called *clean* if for each $z \in Z$ the set of fixed points of $d\Phi_z : T_z M \to T_z M$ equals the tangent space of Z.

REMARK. This definition is due to Bott [3].

If M is a symplectic manifold and Φ a symplectic diffeomorphism, a clean submanifold of fixed points possesses a canonical nowhere vanishing smooth density. To see this we need the following linear algebra lemma:

LEMMA 12.6. *Let* V *be a symplectic vector space with two-form* Ω. *Let* $P: V \to V$ *be a symplectic linear mapping. Then* $\ker(I-P)$ *and* $\operatorname{coker}(I-P)$ *are canonically paired by* Ω.

Proof. If $v \in \ker(I-P)$ then $v = Pv$ so $v = P^{-1}v$, so $v \in \ker(I-P^{-1})$. But to say that $v \in \ker(I-P^{-1})$ is equivalent to saying that $\Omega(v,(I-P)w) = 0$ for all w; or that $v \in \operatorname{Im}(I-P)^{\perp}$. Hence $\ker(I-P)$ and $\operatorname{coker}(I-P)$ are canonically paired. Q.E.D.

Now consider the exact sequence

$$0 \longrightarrow \ker \longrightarrow V \xrightarrow{I-P} V \longrightarrow \operatorname{coker} \longrightarrow 0 .$$

Letting $|\ |^{\alpha}$ be the functor which assigns to each vector space V its one-dimensional space of α-densities we get

$$|(\ker)|^{1/2} \otimes |V|^{-1/2} \otimes |V|^{1/2} \otimes |(\operatorname{coker})|^{-1/2} \cong 1 .$$

Since $|V|^{1/2} \otimes |V|^{-1/2} \cong 1$ and $|(\operatorname{coker})|^{-1/2} \cong |(\ker)|^{1/2}$, we get $|(\ker)| \cong 1$; so we conclude

LEMMA 12.7. *If* P *is a symplectic mapping of* V *into* V *then* $\ker(I-P)$ *possesses a canonical density.*

REMARK. Suppose that P in Lemma 12.7 satisfies

$$(I-P)^{\#} : V/\ker \xrightarrow{\cong} V/\ker , \qquad \ker = \ker(I-P) .$$

Then by Lemma 12.6, Ω restricted to $\ker(I-P)$ is non-singular and hence $\ker(I-P)$ is a symplectic space. It is not hard to see that the density on

$\ker(I-P)$ described by Lemma 12.7 is just the symplectic density times the factor $|\det(I-P^{\#})|^{-1/2}$.

Applying Lemma 12.7 to $Z \subset M$, a clean submanifold of fixed points of a symplectic diffeomorphism, we obtain:

PROPOSITION 12.8. *Z possesses a canonical nowhere vanishing density.*

Let M be a symplectic manifold and $\sigma : M \to \mathbf{R}$ a smooth function with derivative everywhere unequal to zero. Let Ξ be the Hamiltonian vector field associated with σ and $\phi_t : M \to M$ the flow it generates. Let Z' be a submanifold consisting of periodic orbits of Ξ of period t_o. Then Z' is a fixed point set for the symplectic mapping $\phi_{t_o} : M \to M$. Assume that Z' is a clean fixed point set. Then the submanifold

$$Z = Z' \cap (\text{energy surface } \sigma = 1)$$

is a clean fixed point set for the mapping ϕ_{t_o} restricted to the energy surface. Letting μ' be the canonical density on Z' we get a canonical density μ on Z by requiring that at each point of Z, $\mu' = \mu \otimes |d\sigma|$. Now let $M = \Sigma$, and $\sigma = \sigma(T)$. Let $m : Z \to \mathbf{C}$ be the function which, to each point of z, assigns the integral of the subprincipal symbol of T around the periodic orbit of ϕ_t through z. By Proposition 11.10, this function is well defined, at least modulo integer multiples of 2π.

With the notation above, Proposition 12.4 has the following amelioration.

THEOREM 12.9. *Let $\chi_r(t)$ be the distribution $(t+\sqrt{-1}\,0+)^{-r}$. Suppose that the set, Z, of fixed points of the mapping ϕ_{t_o} on $\Sigma \cap (\sigma=1)$ is clean in the sense of Definition 12.5 and that its dimension is d. Then for t near t_o, $e(t)$ admits an asymptotic expansion*

$$e(t) \sim \sum_{r=r_o}^{-\infty} a_r \chi_r(t), \quad r_o = (d+1)/2 . \tag{12.4}$$

The expansion goes down by integral degrees, and the leading term in the expansion is given by the formula

$$(12.5) \qquad a_{r_o} = \gamma_k \int_Z e^{\sqrt{-1}m(z)} d\mu(z) .$$

(γ_k being an innocuous positive constant depending only on k.)

Two special cases of this formula are of particular interest. If Z is one-dimensional, then it consists of a finite set of periodic trajectories of period t_o, say $\gamma_1, \gamma_2, \cdots, \gamma_p$. Let P_i be the linear Poincaré map associated with the trajectory γ_i, and let t_i be the primitive period of γ_i. Theorem 12.9 coupled with the remark following Lemma 12.7 implies

THEOREM 12.10. *Near* t_o, $e(t) = a_1(t - t_o + \sqrt{-1}\, 0+)^{-1} + \dot{e}'(t), e'(t)$ *being a locally* L_1*-summable function. Moreover* a_1 *is given by the formula*

$$(12.6) \qquad a_1 = (1/2\pi)\, |t_i| \sum (1/|I-P_i|^{1/2}) e^{\sqrt{-1} \int_{\gamma_i} \sigma_{sub}(T)} .$$

If $t_o = 0$ then Z is the whole set $\Sigma_1 = \{(x,\xi) \in \Sigma, \sigma(x,\xi) = 1\}$ and from Theorem 12.9 we deduce

THEOREM 12.11. *For* t *near zero*

$$e(t) \sim \sum_{r=k}^{-\infty} a_r \chi_r(t) , \qquad k = (\dim \Sigma)/2$$

with $a_k = k!(2\pi i)^{-k} \operatorname{vol}(\Sigma_1)$.

Proof of Theorem 12.9. As above let $U(x,y,t)$ be the Schwartz kernel of the operator $(\exp \sqrt{-1}\, tQ)\pi$. Let $\mathcal{C}_\Sigma$ be the subset of $T^*(X \times X \times \mathbf{R}) - 0$ consisting of all (x,ξ,y,η,t,τ) for which

a) $(x,\xi) \in \Sigma$,

b) (x,ξ) and $(y,-\eta)$ are the end points of an integral curve of Ξ of length t.

c) $\tau = \sigma(x,\xi)$.

It is easy to see that $\mathcal{C}_\Sigma$ is an isotropic submanifold of $T^*(X\times X\times \mathbf{R}) - 0$.

LEMMA 12.12. $U \in I^{k/2}(X\times X\times \mathbf{R}, \mathcal{C}_\Sigma)$.

Proof. This follows immediately from Theorem 7.5 and results of §5 of [13].

As we observed above the operator, $\pi_*\Delta^*$, in (12.3) is a Fourier integral operator associated with the canonical relation

$$\Gamma = \{((t,\tau),(x,\xi,x,\xi,t,\tau)),(t,\tau) \in T^*\mathbf{R},(x,\xi) \in T^*X\}.$$

Assuming that Γ and $\mathcal{C}_\Sigma$ satisfy the hypotheses (7.4), we can compute $\pi_*\Delta^* U$ by means of Theorem 7.5. Conditions a), b) and c) of (7.4) are trivially satisfied. Identifying Γ with a submanifold of $T^*(X\times X\times \mathbf{R})$ by means of the imbedding $((t,\tau),(x,\xi,x,\xi,t,\tau)) \to (x,\xi,x,\xi,t,\tau)$, condition d) of (7.4) is satisfied providing Γ and $\mathcal{C}_\Sigma$ intersect cleanly in $T^*(X\times X\times \mathbf{R})$. The intersection can be identified with the set of all points (x,ξ,t) such that (x,ξ) is a fixed point of ϕ_t. The intersection is clean if and only if t is locally constant on the intersection (say $t = t_o$) and the intersection is a clean fixed point set for ϕ_{t_o}. (See §4 of [12].) Finally condition e) of (7.4) is satisfied since the rank of the mapping τ in $(7.4)_e$ is everywhere equal to one. We can now apply Theorem 7.5:

LEMMA 12.13. *Let $\Lambda^+_{t_o}$ be the positive component of the tangent space to $\mathbf{R}$ at t_o. If the hypotheses of Theorem 12.9 are satisfied then near $t = t_o$, $e(t)$ is in $I^{(d-1)/2}(\mathbf{R},\Lambda^+_{t_o})$.*

By [18], §6.4, $I^{(d-1)/2}(\mathbf{R},\Lambda^+_{t_o})$ consists precisely of those distributions which admit an asymptotic expansion of the form (12.4); so this proves the first half of Theorem 12.9. The fact that the terms in (12.4) go down by *integral* rather than half-integral degrees is a consequence of Theorem 9.9 and the remarks following it. To compute the coefficient, a_{r_o}, of the leading term in (12.4) we must compute the leading symbol of

$\pi_* \Delta^* U$ using the recipe described in §7. This involves first of all looking carefully at the symbol of U. By definition $(\exp \sqrt{-1}\, tQ)\pi$ satisfies the operator equation

$$((1/\sqrt{-1})(\partial/\partial t) - Q)(\exp \sqrt{-1}\, tQ)\pi = 0$$

with initial data given by π at $t = 0$; so the symbol of U must satisfy the transport equation (10.3) on $\mathcal{C}_\Sigma$ with pre-determined initial data. It is clear that this determines the symbol completely.

To describe the symbol it will be useful to describe separately the half-form component of the symbol and the "spin" component. The half-form component being easiest to describe we will discuss it first. Let $\nu = \sqrt{\Omega^d}$ be the canonical half-form on Σ associated with the symplectic form Ω. Let $p = ((x,\xi), \phi_{t_o}(x,\xi), t_o, \tau)$ be a point of $\mathcal{C}_\Sigma$ and let $\gamma(t)$, $0 \le t \le t_o$, be the integral curve of Ξ joining (x,ξ) to $\phi_{t_o}(x,\xi)$. The half-form component of the symbol of U takes the following value at p:

$$e^{\sqrt{-1}\int_o^{t_o} \sigma_{\mathrm{sub}}(Q)\, dt} (\phi_{t_o})_* \nu \otimes \nu . \tag{12.7}$$

Note that (12.7) satisfies the transport equation and at $t = 0$ is just $\nu \otimes \nu$. (Compare with §4 of [12].)

Next we will describe the spin component of the symbol. Given $z = (x,\xi) \in \Sigma$, let $\mathcal{S}_z$ be the spin space at z and $e_\Lambda(z) \in \mathcal{S}_z$ the "vacuum state" determined by Λ_z. We showed in §11 that the spin component of the symbol of π at z is the projection operator $\pi_\Lambda : \mathcal{S}_z \to \mathcal{S}_z$ projecting $\mathcal{S}_z$ onto the one-dimensional space spanned by $e_\Lambda(z)$. Let $z' = \phi_t(z)$. By Proposition 11.5 $\phi_t e_\Lambda(z)$ is (up to multiples of modulus one) the vacuum state associated with Λ in $\mathcal{S}_{z'}$. Let $U_\Lambda(z,t)$ be the operator from $\mathcal{S}_z$ to $\mathcal{S}_{z'}$ which maps $e_\Lambda(z)$ into $\phi_t(e_\Lambda(z))$ and maps the orthogonal complement of $e_\Lambda(z)$ onto zero.

LEMMA 12.14. *The symbol of* U *at* (z, z', t) *is the product of the half-form (12.7) and* U_Λ.

Proof. It is easy to check that this candidate for the symbol satisfies the transport equation and takes on the appropriate value at $t = 0$. Q.E.D.

We must now compute the leading symbol of $\pi_* \Delta^* U$. The computation is nearly identical with the analogous computation in §4 of [12]; so we will just indicate the main steps. Let $z = (x, \xi)$ be a point of Σ such that the integral curve, γ, of Ξ through z is periodic of period t_o; and let (z, z, t_o, τ) be the corresponding point on the intersection of $\mathcal{C}_\Sigma$ with Γ. Let $e_\Lambda : \gamma \to SV$ be a trivialization of the "vacuum state" bundle along γ. Then $(1/\sqrt{-1})\Xi e_\Lambda = c(Q) e_\Lambda$. Integrating this equation along γ we get

$$(\phi_t)_* e_\Lambda = e^{\sqrt{-1}\int_o^t c(Q)\,dt} e_\Lambda .$$

In particular for $t = t_o$ we get

$$U_\Lambda = e^{\sqrt{-1}\int_\gamma c(Q)} e_\Lambda \otimes e_\Lambda .$$

Combining this result with Lemma 12.14 we get

LEMMA 12.15. *At a point* (z, z, t_o, τ) *on the intersection of* $\mathcal{C}_\Sigma$ *with* Γ *the symbol of* U *is given by the formula*

$$(e^{\sqrt{-1}\int_\gamma \sigma_{\text{sub}}(T)} \pi_\Lambda) \otimes \nu \otimes (\phi_t)_* \nu . \tag{12.8}$$

The trace of the operator π_Λ is just one, and by means of Lemma 12.7, $\nu \otimes (\phi_t)_* \nu$ can be canonically identified with the density μ on Z. Taking the trace of the first factor in (12.8), replacing the second factor by μ and integrating over Z, we get (12.5) for the leading symbol of $\pi_* \Delta^* U$.

§13. SPECTRAL PROPERTIES OF TOEPLITZ OPERATORS

Let X be a compact metalinear manifold and Σ a closed homogeneous symplectic submanifold of T^*X-0 of dimension $2n$. Let there be given a Toeplitz structure on Σ and let T be a self-adjoint first order Toeplitz operator with $\sigma(T)>0$ everywhere. Let $\lambda_1 \leq \lambda_2 \leq \lambda_3 \leq \cdots$ be its spectrum.

THEOREM 13.1. *If* $N(\lambda)$ *is the number of* λ_i*'s less than* λ *then*

$$N(\lambda) = (\mathrm{vol}(\Sigma_1)/(2\pi)^n)\lambda^n + O(\lambda^{n-1}), \tag{13.1}$$

Σ_1 *being the subset,* $\sigma(T) \leq 1$, *of* Σ *and* $\mathrm{vol}(\Sigma_1)$ *its symplectic volume.*

Proof. Let $e(t) = \sum e^{\sqrt{-1}\,\lambda_i t}$ and let $\rho: \mathbf{R} \to \mathbf{R}$ be a smooth function with support in a neighborhood of the origin on which the asymptotic expansion in Theorem 12.11 is valid. Using the well-known formula

$$e^{\sqrt{-1}(\pi r/2)}\Gamma(r)(t+io)^{-r} = \int_0^\infty e^{\sqrt{-1}\,st} s^{r-1}\,ds$$

we get from Theorem 12.11,

$$\rho(t)e(t) = (\mathrm{vol}(\Sigma_1)/(2\pi)^n)\int_{-\infty}^{\infty} e^{\sqrt{-1}\,st}(ns_+^{n-1}+h(s))\,ds$$

$h(s)$ being of order $O(s^{n-2})$ for $s >> 0$. Taking the Fourier transform of both sides of this equation we get

$$\sum \hat{\rho}(s-\lambda_i) = (\mathrm{vol}(\Sigma_1)/(2\pi)^n)(ns^{n-1}) + O(s^{n-2}) \tag{13.2}$$

for $s >> 0$. From (13.2) one deduces (13.1) by an argument which is by now fairly standard. (See §6 of [24].) Q.E.D.

Let Ξ be the Hamiltonian vector field on Σ associated with the function $\sigma = \sigma(T)$. The following result is due to Helton, [21].

THEOREM 13.2. *Let* Δ *be the set of cluster points of the set* $\{\lambda_i - \lambda_j\}$. *Then* $\Delta \neq R$ *only if the trajectory of* Ξ *through every point of* X *is periodic.*

The proof depends upon the following Egorov theorem for Toeplitz operators:

PROPOSITION 13.3. *Let* T *be as above and let* A *be an arbitrary zeroth order Toeplitz operator. Let* $A_s = (\exp\sqrt{-1}\, sT)A(\exp\sqrt{-1}(-s)T)$. *Then* A_s *is again a zeroth order Toeplitz operator and* $\sigma(A_s) = (\exp s\Xi)^*\sigma(A)$.

Proof. Let $T = \pi Q \pi$, Q being a self-adjoint first order pseudodifferential operator with $\sigma(Q) > 0$ and $[Q, \pi] = 0$. Let $A = \pi B \pi$, B being a zeroth order pseudodifferential operator. Then if $B_s = (\exp\sqrt{-1}\, sQ)B(\exp\sqrt{-1}(-s)Q)$, $A_s = \pi B_s \pi$; so the theorem above follows from the usual Egorov theorem for pseudodifferential operators. Q.E.D.

The proof of Theorem 13.2, based on the Egorov theorem, is identical with the proof given in [21]. We refer to chapter one of [16] for details. Q.E.D.

Suppose that all trajectories of Ξ are periodic of period τ. We will refine Theorem 13.2 by first of all proving:

PROPOSITION 13.4. *The operator* $\exp\sqrt{-1}\,\tau T$ *is a Toeplitz operator of order zero. Moreover the value of its leading symbol at* $(x,\xi) \in \Sigma$ *is*

$$\exp\sqrt{-1}\int_\gamma \sigma_{sub}(T) \tag{13.3}$$

σ_{sub} *being the subprincipal symbol of* T *and* γ *the trajectory of* Ξ *through* (x,ξ).

Proof. Let $U(t) = \exp\sqrt{-1}\,tT$. We showed in §12 that $U(t)$ is a Fourier integral operator of Hermite type with wave front set supported on the graph of the canonical transformation $\exp t\Xi : \Sigma \to \Sigma$. If all trajectories of Ξ are periodic of period τ then $\exp \tau\Xi$ is the identity map; so the wave front set of $U(\tau)$ is concentrated on the diagonal in $\Sigma\times\Sigma$. This shows that $U(\tau) \in OP\mathcal{H}^{0}(\Sigma)$. Since $U(\tau) = \pi U(\tau)\pi$, the leading symbol of $U(\tau)$ is a multiple of π_Λ by Theorem 9.8. This means we can find a classical pseudodifferential operator, B, of order zero such that $U(\tau) - \pi B\pi \in OP\mathcal{H}^{-1/2}(\Sigma)$. In fact, by parity considerations, $U(\tau) - \pi B\pi \in OP\mathcal{H}^{-1}(\Sigma)$ it is easy to see that both the Schwartz kernel of $U(\tau)$ and the Schwartz kernel of $\pi B\pi$ are Hermite distributions of *even* type; so the symbol of $U(\tau) - \pi B\pi$ must be an *odd* symplectic spinor. However $U(\tau) - \pi B\pi = \pi(U(\tau) - \pi B\pi)\pi$; so, as above, this symbol is a multiple of π_Λ, which is an even symplectic spinor. It follows that the symbol of order $-\frac{1}{2}$ of $U(\tau) - \pi B\pi$ vanishes; so $U(\tau) - \pi B\pi \in OP\mathcal{H}^{-1}(\Sigma)$.

The same argument as before shows that there exists a pseudodifferential operator, B_1, of order -1 such that $U(\tau) - \pi(B_0 + B_1)\pi$ is of order -2. Inductively we can construct a sequence $B_0, B_1, B_2, \cdots$ such that $U(\tau) - \pi(B_0 + \cdots + B_N)\pi$ is in $OP\mathcal{H}^{-N-1}(\Sigma)$. Let B' be a pseudodifferential operator such that for all N, $B' - (B_1 + \cdots + B_N)$ is of order $-(N+1)$. Then $U(\tau) - \pi B\pi$ is smoothing. This proves that $U(\tau)$ is a Toeplitz operator. Finally by Lemma 12.14 the leading symbol of $U(\tau)$ is the symbol (13.2). Q.E.D.

We will use this result to prove

THEOREM 13.5. *Let* Δ' *be the cluster set of the set of points* $\{e^{\sqrt{-1}\,\lambda_i\tau}\}$. *Then the following are equivalent*

a) $\operatorname{Int}\Delta' = \emptyset$ *in* S^1.

b) Δ' *consists of a single point.*

c) *The integral of* $\sigma_{sub}(T)$ *over any periodic trajectory,* γ, *of* Ξ *is a constant, independent of* γ.

For the proof we will need the following two lemmas.

LEMMA 13.6. *Let* S *be a self-adjoint zeroth order Toeplitz operator and let* f *be a smooth real-valued function on the real line. Then the operator* f(S) *is a self-adjoint zeroth order Toeplitz operator and* $\sigma(f(S)) = f(\sigma(S))$.

Proof. Let $S = \pi Q \pi$, Q being a self-adjoint zeroth order pseudodifferential operator such that $[Q, \pi] = 0$. Then $f(S) = \pi f(Q) \pi$, and the theorem above follows from an analogous theorem for f(Q) proved in [19]. Q.E.D.

LEMMA 13.7. *Let* S *be a self-adjoint zeroth order Toeplitz operator whose spectrum is countable (counting multiplicities) and consists of the points* $\{\mu_i\}$. *Let* Γ *be the cluster set of the set* $\{\mu_i\}$. *Then* $\Gamma = [a, \beta]$ *where* $a = \min \sigma(S)$ *and* $\beta = \max \sigma(S)$.

Proof. Let f(t) be a real-valued function supported in the interval $\beta < t < \infty$. By Lemma 13.6, the leading symbol of f(S) vanishes; so f(S) is a Toeplitz operator of order -1. In particular f(S) is compact; so its spectrum $\{f(\mu_i)\}$ tends to zero as i goes to ∞. If f is a function which is equal to one on the interval $t > \beta + \varepsilon$, $\varepsilon > 0$, this shows that only finitely many μ_i's lie above $\beta + \varepsilon$. A similar argument shows that only finitely many μ_i's lie below $a - \varepsilon$; so Γ is contained in $[a, \beta]$. To show that $\Gamma = [a, \beta]$, suppose there exists $\mu_o \in [a, \beta]$ such that $\mu_o \notin \Gamma$. Then there exists a $\delta > 0$ such that only a finite number of μ_i's lie in the interval $(\mu_o - \delta, \mu + \delta)$. Let f be a smooth function supported on this interval. Then on the one hand f(Q) is an operator of finite rank, while on the other hand $\sigma(f(Q)) = f(\sigma(Q)) \neq 0$. This contradiction proves the theorem. Q.E.D.

To prove Theorem 13.5 we apply Lemma 13.7 to the operator $\sin \tau T$. The symbol of $\sin \tau T$ is equal to $\sin \int_\gamma \sigma_{sub}(T)$ at (x, ξ), where γ

is the integral curve of Ξ through (x,ξ); so either $\int_\gamma \sigma_{\mathrm{sub}}(T)$ is constant or the cluster set of $\{\sin \tau\lambda_i\}$ contains an open interval. If the latter happens, then the cluster set of $\{e^{\sqrt{-1}\,\lambda_i \tau}\}$ contains an open interval. Furthermore, Lemma 13.7 says that if $\int_\gamma \sigma_{\mathrm{sub}}(T)$ is constant, the cluster set consists of a single point. Q.E.D.

Suppose now that any (and hence all) of the three conditions, of Theorem 3.5 is satisfied. Let m_o be the constant value of $\int_\gamma \sigma_{\mathrm{sub}}(T)$. By Proposition 3.4, $\exp \sqrt{-1}\,\tau T - (e^{\sqrt{-1}\,m_o})\pi$ is a Toeplitz operator of order -1. Replacing T by $(2\pi)^{-1}(\tau T - m_o)$, which has the same spectrum as T up to an affine transformation of the line, we will henceforth assume that $\exp 2\pi\sqrt{-1}T - \pi$ is a Toeplitz operator of order -1. Following Colin de Verdiére, [10], we will prove

THEOREM 13.8. *There exist Toeplitz operators,* U *and* V, *such that* $T = U + V$, V *is of order* -1, U *and* V *commute and* $\exp 2\pi\sqrt{-1}\,U = \pi$.

Proof. Let $\phi_1, \phi_2, \cdots$ be normalized eigenfunctions for T corresponding to the eigenvalues $\lambda_1, \lambda_2, \cdots$. Let $C = \exp 2\pi\sqrt{-1}\,T - \pi$. The ϕ_i's are normalized eigenfunctions for C corresponding to the eigenvalues $c_i = \exp 2\pi\sqrt{-1}\,\lambda_i - 1$. Since C is of order -1, C is compact and the c_i's tend to zero. Let v_n be a number such that $v_n = (1/2\pi\sqrt{-1})\log(1+c_n)$. (For n sufficiently large, say $n > n_o$, v_n is determined uniquely by specifying it to be the principal value of $\log(1+c_n)$. For $n < n_o$ let v_n be any value of $\log(1+c_n)$.) Let V be the operator defined by $V\phi_i = v_i\phi_i$, $i = 1, 2, \cdots$. Clearly $[V, T] = 0$ and if $U = T - V$, $\exp 2\pi\sqrt{-1}\,U = \pi$. It remains to show that V is a Toeplitz operator of order -1. Let C' be defined by

$$C'\phi_n = 0 \quad \text{for} \quad n \le n_o \quad \text{and} \quad C'\phi_n = c_n\phi_n \quad \text{for} \quad n > n_o .$$

Then

$$V = (1/2\pi\sqrt{-1}) \sum_{k=1}^{\infty} (-1)^k((C')^k/k) + W$$

W being a finite rank smoothing operator and the kth term on the right a Toeplitz operator of order $-k$. Q.E.D.

An immediate corollary of Theorem 13.8 is the following

THEOREM 13.9. *There exists a constant* $M > 0$ *and an integer* k_o *such that all but finitely many eigenvalues of* T *lie on the union of the intervals*

$$(13.4)_k \qquad |\lambda - k| \leq M/k\,, \quad k \geq k_o\,.$$

Our next two results concern the multiplicity and distribution of the eigenvalues on the interval $(13.4)_k$. To simplify the statement of these results we will henceforth assume that the trajectories of Ξ are *simply* periodic of period 2π. (This assumption can be somewhat weakened, as Colin has shown in the pseudodifferential case in [10].)

PROPOSITION 13.10. *There exists a polynomial* $p(t)$ *of degree* $n-1$ *such that for* $k >> 0$, *the number of eigenvalues of* T *in the interval* $(13.4)_k$ *is* $p(k)$.

Proof. Since the number of eigenvalues of T on the interval $(13.4)_k$ is equal to the multiplicity of the k-th eigenvalue of U it is enough to prove the theorem with T replaced by U. Consider the one parameter group $\exp \sqrt{-1}\, tU$. This is periodic of period 2π; so by Proposition 12.3 trace $\exp \sqrt{-1}\, tU$ is a well-defined distribution on S^1. Consider the distribution

$$(13.6)_r \qquad \theta_r(t) = \sum_{k=1}^{\infty} k^{r-1} e^{ikt}$$

on S^1. θ_r and χ_r have identical singularities at $t = 0$; so by Proposition 12.11 trace $\exp \sqrt{-1}\, tU$ admits an asymptotic expansion

$$(13.7) \qquad \text{trace} \exp \sqrt{-1}\, tU \sim \sum_{r=n}^{-\infty} a_r \theta_r(t)$$

near $t = 0$. Moreover, since the trajectories of Ξ are *simply* periodic of period 2π, trace $\exp\sqrt{-1}\, tU$ is smooth except at the points $\{2\pi m, m \in Z\}$ by Proposition 12.4. Therefore the asymptotic expansion (13.7) is valid on the whole of S^1. On the other hand if d_k is the dimension of the k-th eigenspace of U we have

$$\text{trace} \exp \sqrt{-1}\, tU = \sum d_k e^{ikt} . \tag{13.8}$$

Comparing the k-th Fourier coefficients of (13.7) and (13.8) we get

$$d_k \sim \sum_{r=n-1}^{-\infty} a_r k^r . \tag{13.9}$$

However, the d_k's are *integers*. A simple arithmetic argument (which we leave as an exercise for the reader) shows that if an asymptotic expansion of the form (13.9) is valid and the left-hand side takes *integer* values, then $a_r = 0$ for $r < 0$ and for k sufficiently large, $d_k = \sum_{r=n-1}^{0} a_r k^r$. Q.E.D.

REMARK. This proof is due to Colin de Verdiére (see [10]). An explicit formula for $p(t)$ will be given in the next section.

The next theorem is a Toeplitz operator analogue of a Szegö type theorem for pseudodifferential operators due to Weinstein [35].

THEOREM 13.11. *Let M be a self-adjoint zeroth order Toeplitz operator which commutes with T. Let M_k be the induced operator on the space of eigenfunctions associated with the eigenvalues in the cluster $(13.4)_k$. Let $\lambda_i^{(k)}$, $i = 1, \cdots, d_k$ be the eigenvalues of M_k and let μ_k be the measure*

$$\mu_k = (1/k^{n-1}) \sum \delta(\lambda - \lambda_i^{(k)}) .$$

Then as $k \to \infty$ the measure μ_k tends weakly to a limit μ, μ being given by the formula

$$(13.10) \qquad \mu(f) = \gamma_n \int_{\sigma(T)=1} f(\sigma(M))(z)\,dz \quad \textit{for} \quad f \in C_o(R)\,.$$

Here γ_n *is a universal constant depending only on* n *and* dz *is the measure induced on the subset,* $\sigma(T) = 1$, *of* Σ *by the symplectic measure on* Σ.

REMARK. Applying this theorem to UV, V being as in Theorem 13.8, we get a picture of the asymptotic distribution of the eigenvalues of T itself on the interval $(13.4)_k$ as k tends to infinity.

Proof of Theorem 13.11. Let f be a smooth function on the real line. By Lemma 13.6, f(M) is a zeroth order Toeplitz operator whose leading symbol is $f(\sigma(M))$. Let U and V be as in Proposition 13.8. The k-th eigenspace of U is identical with the space spanned by the eigenfunctions of T with eigenvalues lying on the interval $(13.4)_k$; so U and M commute and

$$(13.11) \qquad \text{trace } f(M) \exp \sqrt{-1}\,tU = \sum k^{n-1}\mu_k(f)e^{ikt}\,.$$

On the other hand, one can show that there is an asymptotic expansion

$$(13.12) \qquad \text{trace } f(M) \exp \sqrt{-1}\,tU \sim \sum_{r=n}^{-\infty} a_r(f)\theta_r(t)$$

with

$$a_n = \gamma_n \int_{\sigma(T)=1} \sigma(f(M))(z)\,dz\,.$$

(The proof of (13.11) is, with small modifications, identical with the proof of (13.7); so we will omit it.) Comparing coefficients in (13.11) and (13.12) we get an asymptotic expansion of the form

$$(13.13) \qquad \mu_k(f) \sim \sum_{r=n}^{-\infty} a_r(f)k^{r-m}\,.$$

The leading term in this expansion gives us (13.10). Q.E.D.

We will conclude this section by describing a curious application of Theorem 13.11 to complex variable theory. Let L be the canonical line bundle over complex projective n-space. If we think of CP^n as the set of all one-dimensional subspace of C^{n+1} then the fiber, L_p, at $p \in CP^n$ is the one-dimensional subspace of C^{n+1} represented by p. If we endow C^{n+1} with its standard Hermitian inner product, $(z, w) \to z\bar{w}$, then each of the L_p's acquires an inner product; so L is a Hermitian line bundle. Let V be a non-singular, r-dimensional, projective subvariety of CP^n and let $L(V)$ be the restriction of L to V. The following fact is well known.

PROPOSITION 13.12. *Let* W *be the set of all pairs* $\{(p, v), p \in V, v \in L_p, |v| \leq 1\}$. *Then* W *is a strictly pseudoconvex domain.*

The boundary of W is the set

$$\partial W = \{(p, v), p \in V, v \in L_p, |v| = 1\} ;$$

so it is a circle bundle over V. Its canonical contact form, α, can be viewed as a connection form on this circle bundle and the corresponding curvature form is a symplectic two-form on V. This shows that the Hermitian structure on L produces a canonical symplectic structure on V. Let ν be the volume form on V associated with this symplectic structure. From ν and from the inner product on $L(V)$ we get a pre-Hilbert structure on the space of smooth sections of $L(V)$. The same is true for the space of smooth sections of each of the line bundles $\overset{k}{\otimes} L^*(V)$, $k = 0, 1, 2, \cdots$. For k large, the line bundles $\overset{k}{\otimes} L^*(V)$ admit lots of global holomorphic sections. Indeed let S^k be the space of all homogeneous polynomial functions of degree k on C^{n+1}, let I^k be the subspace of those polynomials which vanish on V, and let Γ^k be the space of holomorphic sections of $\overset{k}{\otimes} L^*$. Then $\Gamma^k \cong S^k/I^k$.

Now let ρ be a smooth real-valued function on V. We can define an operator

$$M_\rho^k : \Gamma^k \to \Gamma^k \tag{13.14}$$

as follows. If s is a holomorphic section of $\overset{k}{\otimes} L^*$ we multiply it by ρ to get a smooth section, ρs, of $\overset{k}{\otimes} L^*$. Thanks to the pre-Hilbert structure we can project ρs back into Γ^k. We will denote this projection by $M_\rho^k s$. Since ρ is real-valued, M_ρ^k is a self-adjoint linear mapping. Let

$$\lambda_i^{(k)}, \quad i = 1, \cdots, d_k$$

be its eigenvalues, (here $d_k = \dim \Gamma^k$) and let μ_k be the measure

$$\mu_k = 1/k^r \sum_{i=1}^{d_k} \delta(\lambda - \lambda_i^{(k)}) .$$

THEOREM 13.13. *As k tends to infinity μ_k tends weakly to a limiting measure, μ, which is given explicitly by the formula*

$$\mu(f) = \gamma_r \int_V f(\rho(z))\, d\nu(z) \quad \textit{for} \quad f \in C_0(R) .$$

(γ_r is a universal constant depending only on r.)

Proof. Let $\partial/\partial\theta$ be the generator of the circle group action on ∂W. $\partial/\partial\theta$ preserves the boundary complex structure; so if π is the Szegö projector, π and $\partial/\partial\theta$ commute. The restriction of $(1/\sqrt{-1})\,\partial/\partial\theta$ to the Hardy space is a first order self-adjoint Toeplitz operator which we will denote by T. Since $\alpha(\partial/\partial\theta) = 1$, $\sigma(T) > 0$ everywhere, and the contact vector field on ∂W associated with $\sigma(T)$ is just $\partial/\partial\theta$.

LEMMA 13.14. *The spectrum of T consists of the non-negative integers and the k-th eigenspace of T is identical with Γ^k.*

Proof. Given $p \in V$, let z be a complex coordinate function on the complex line L_p. The fiber of ∂W at p can be identified with the circle

$|z| = 1$. Let f be a k-th eigenfunction of T and let $\tilde{f}$ be its restriction to ∂W_p. Then $(1/\sqrt{-1})\partial/\partial\theta\, \tilde{f} = k\tilde{f}$; so $\tilde{f} = az^k$, a being a constant which depends smoothly on p. This shows that f can be regarded as a smooth section of $\overset{k}{\otimes} L^*$. The fact that f satisfies the boundary Cauchy-Riemann equations simply says that f is a holomorphic section of $\overset{k}{\otimes} L^*$.

Q.E.D.

We will now prove Theorem 13.13. Let $\tilde{\rho}$ be the pull-back of ρ to ∂W and let M_ρ be the zeroth order Toeplitz operator sending an element, f, of the Hardy space onto $\pi\tilde{\rho}f$, π being the Szegö projector as above. Since $\partial/\partial\theta\,\tilde{\rho} = 0$, T and M_ρ commute; and M_ρ restricted to the k-th eigenspace of T is just M_ρ^k by Lemma 13.14. Now apply Theorem 13.11.

Q.E.D.

For further remarks concerning this "complex Szegö theorem" we refer to [17].

§14. THE HILBERT POLYNOMIAL

Let M be a compact 2r-dimensional symplectic manifold whose symplectic two-form determines an *integral* cohomology class in $H^2(M, R)$. We are going to associate with M a degree r polynomial of one variable, $p(x)$, which we will call the *Hilbert polynomial* of M. It will have the property that whenever n is an integer, $p(n)$ is an integer. We will give two alternative definitions of this polynomial, one topological and one analytic. We will begin with the topological definition.

Let V be a 2r-dimensional real symplectic vector space with symplectic form, ω. Let V also be equipped with an inner product, $\langle\ ,\ \rangle$. Then there exists a non-singular mapping $B : V \to V$ such that for all $v, w \in V$, $\langle Bv, w\rangle = \omega(v, w)$. Clearly $B = -B^t$; so the spectrum of B lies entirely on the imaginary axis in the complex plane. Since B is real, exactly half of its eigenvalues lie on the positive imaginary axis. Let $\Lambda^{1,0}$ be the subspace of $V \otimes C$ spanned by the eigenvectors associated with these eigenvalues. Given $v \in V$ there exists a unique vector $Jv \in V$ such that $v + \sqrt{-1}\, Jv$ is in $\Lambda^{1,0}$. Clearly $J^2 = -1$; so J defines a complex structure on V.

More generally, if M is a symplectic manifold, let $\langle\ \rangle$ be an inner product on its tangent bundle. Then $\langle\ \rangle$ defines a complex structure on the tangent bundle, or, what amounts to the same thing, an almost complex structure on M. Given two inner products, $\langle\ \rangle_0$ and $\langle\ \rangle_1$, let $\langle\ \rangle_t = (1-t)\langle\ \rangle_0 + t\langle\ \rangle_1$. This is also an inner product; so *any two almost complex structures on* M *defined in this way are homotopic*. In particular the Todd class $\tau \in H^*(M, Z)$ associated with any one of these structures is identical with the Todd class associated with any other. (For the definition of the Todd class see [22].) We will henceforth call τ

the *Todd class* of M. Suppose now that the symplectic form, ω, determines an integral cohomology class in $H^2(M, \mathbf{R})$.

Then

$$p_{top}(x) = (e^{-x[\omega]}\tau)[M] \tag{14.1}$$

takes on integer values for integer values of x. We will call (14.1) the *topological* Hilbert polynomial of M.

Next we will define a polynomial, $p_a(x)$, which we will call the *analytic* Hilbert polynomial of M. To simplify the definition somewhat we will assume M is simply connected (though this assumption can be avoided with a little extra work). The definition of the polynomial, p_a, is unfortunately rather circuitous. Suppose that, instead of M, we are given the following data: a compact metalinear manifold, X, a closed, homogeneous symplectic submanifold, Σ, of $T^*X - 0$, and a smooth homogeneous function of degree one, $\Phi : \Sigma \to \mathbf{R}^+$ having the property that if $\Xi = \Xi_\Phi$ is its Hamiltonian vector field, then the trajectories of Ξ are all simply periodic of period 2π. Let Σ_1 be the subset, $\Phi = 1$, of Σ. The map, $\theta \in S^1 \to \exp \theta\Xi$, defines a free circle group action on Σ_1. The orbit space is a compact Hausdorff manifold, M, of dimension two less than Σ. Moreover $\Sigma_1 \to M$ is a principal circle bundle over M. Let α_1 be the restriction to Σ_1 of the canonical one-form on T^*X. It is easy to see that α_1 is a connection form for the principal fibration $\Sigma_1 \to M$ and that its curvature form, ω, defines a symplectic structure on M. Moreover, since ω is a curvature form, its cohomology class is integral.

Now suppose we are given a Toeplitz structure on Σ. We will prove

PROPOSITION 14.1. *There exists a self-adjoint Toeplitz operator* T *such that* $\sigma(T) = \Phi$ *and such that the spectrum of* T *lies on the union of the intervals* $(13.4)_k$.

Proof. Let T' be any self-adjoint Toeplitz operator with $\sigma(T') = \Phi$. According to §11, $\exp \sqrt{-1} \int_\gamma \sigma_{sub}(T')$ is well defined as a function on the set of trajectories, γ, of Ξ; in other words, it is well defined as an S^1-valued function on M. Call this function f. The real-valued one-form $(2\pi i)^{-1} d(\log) f$ is closed; so if $\pi_1(M) < \infty$ there exists a real-valued function, g, such that $f = \exp(-\sqrt{-1}\, 2\pi g)$. Let T'' be a zeroth order Toeplitz operator such that $\sigma(T'')$ is equal to the pull-back of g to Σ_1, and let $T = T' + T''$. Then for every trajectory, γ, $\exp \sqrt{-1} \int_\gamma \sigma_{\text{sub}}(T) = 1$; so by Theorem 13.9, the spectrum of T lies on the union of the intervals $(13.4)_k$. Q.E.D.

PROPOSITION 14.2. *Let T_0 and T_1 be self-adjoint first order Toeplitz operators such that $\sigma(T_0) = \sigma(T_1) = \Phi$ and such that both the spectrum of T_0 and the spectrum of T_1 lie on the union of the intervals $(13.4)_k$. Then there exists an integer k_0 such that for all k sufficiently large, the number of eigenvalues of T_0 on the k-th interval (13.4) is equal to the number of eigenvalues of T_1 on the $k+k_0$-th interval.*

Proof. Let $V = T_1 - T_0$. Then $\int_\gamma V$ is a constant function on the set of trajectories, γ, of Ξ taking a fixed integer value, k_0. Replacing T_0 by $T_0 + k_0 I$, we can assume this integer is zero. Let $T_s = (1-s) T_0 + s T_1$. Then $\sigma(T_s) = \Phi$ and $\int_\gamma \sigma_{\text{sub}}(T_s)$ is a constant independent of γ; so for M sufficiently large the spectrum of T_s lies on the union of the intervals $(13.4)_k$. However, for fixed k, the number of eigenvalues on the interval $(13.4)_k$ is independent of s. Q.E.D.

Let T be as in Proposition 14.1. By Proposition 13.10, there exists a polynomial, $p(x)$, of degree $(\dim M)/2$ such that the number of eigenvalues of T on the interval $(13.4)_k$ is equal to $p(k)$ for k large. By Proposition 14.2 p is determined (up to translation by an integer, k_0) by Σ, Φ and the given Toeplitz structure on Σ. We can get rid of the ambiguity caused by k_0 by requiring, for example, that $p(z) = 0$ have distinct roots $z_1, \cdots, z_d$ with $\operatorname{Re} z_1 \leq \cdots \leq \operatorname{Re} z_d$ and $0 \leq \operatorname{Re} z_1 < 1$. We will call p, so normalized, the *Hilbert polynomial* of the pair Σ, Φ. To justify this terminology we will prove that the definition of p is independent of the choice of the Toeplitz structure. First we will show:

PROPOSITION 14.3. *Suppose we are given two Toeplitz structures on* Σ. *Let* π *and* π_1 *be the Szegö projectors associated with these structures. Suppose* $\sigma(\pi) = \sigma(\pi_1)$. *Then the Hilbert polynomial defined in terms of* Σ, Φ *and* π *is identical with the Hilbert polynomial defined in terms of* Σ, Φ *and* π_1.

For the proof we will need

LEMMA 14.4. *Let* S *be in* $OP\mathcal{H}^m(\Sigma)$. *Then* $\pi S \pi$ *is a Toeplitz operator of order* m.

Proof. If S is in $OP\mathcal{H}^m(\Sigma)$, $\pi S\pi$ is in $OP\mathcal{H}^m(\Sigma)$ by Theorem 9.8 and its leading symbol is a multiple of $\sigma(\pi)$. This means we can find a pseudodifferential operator, Q_0, of order m such that $\pi S\pi - \pi Q_0 \pi$ is in $OP\mathcal{H}^{m-\frac{1}{2}}(\Sigma)$. The same argument applied to this difference shows that there exists a pseudodifferential operator Q_1 of order $m - \frac{1}{2}$ such that $\pi S\pi - \pi(Q_0 + Q_1)\pi$ is in $OP\mathcal{H}^{m-1}(\Sigma)$. Inductively there exists a sequence of operators, Q_i, of order $m - i/2$ such that $\pi S\pi - \pi(Q_0 + \cdots + Q_N)\pi$ is in $OP\mathcal{H}^{m-N/2}(\Sigma)$. Choose Q so that $Q - (Q_0 + \cdots + Q_{N-1})$ is a pseudodifferential operator of order $m - N/2$. Then $\pi S\pi - \pi Q\pi$ is smoothing.

Q.E.D.

REMARK. If m is an integer and S is in $OP\mathcal{H}_{\text{even}}$ then by parity considerations the Q_i's above can all be chosen to be of integer order of homogeneity. (See the proof of Proposition 13.4.)

We will now prove Proposition 14.3. Let $U = \pi_1 \pi$. Then $U \in OP\mathcal{H}^0(\Sigma)$ and $\sigma(U) = \sigma(\pi_1\pi) = \sigma(\pi)$ by Theorem 9.8. By the lemma there exists a zeroth order pseudodifferential operator Q such that $\sigma(Q) = 1$ and $\pi U = \pi Q \pi$. We can also assume that $[Q, \pi] = 0$, $Q^* = Q$ and $Q \geq 0$. Then there exists a unique self-adjoint pseudodifferential operator, P, of order zero such that P^2 is the "Green's operator" associated with Q, i.e. $P^2Q - I$ is orthogonal projection onto the kernel of Q. Clearly P also commutes with π. Let $V = UP$. Then $V^*V = PU^*UP = \pi PQP\pi = \pi + R$, R being orthogonal projection onto a finite dimensional subspace of the range of π. Now let T be a self-adjoint Toeplitz operator associated with the projector π. Then $T_1 = VTV^*$ is a self-adjoint Toeplitz operator associated with the projector π_1, and the symbol of T is the same as the symbol of T_1. Replacing T by $(I-R)T(I-R)$, which differs from T by a smoothing operator, we can assume that $RT = TR = 0$. Then the spectrum of T_1 is identical with the spectrum of T. In particular if T has its spectrum concentrated on the union of the intervals $(13.4)_k$, so does T_1; and the number of eigenvalues on this interval is the same for T and T_1. Q.E.D.

Next we will show that the Hilbert polynomial does not depend on the Toeplitz structure even when $\sigma(\pi) \neq \sigma(\pi_1)$. Let N be the conormal bundle of Σ and let Λ and Λ_1 be positive definite Lagrangian subbundles of $N \otimes C$ such that $\sigma(\pi) = \pi_\Lambda$ and $\sigma(\pi_1) = \pi_{\Lambda_1}$. (See §11.) To prove that the Hilbert polynomial does not depend on $\sigma(\pi)$ we will need

LEMMA 14.5. *There exist conic neighborhoods,* $\mathcal{C}$ *and* $\mathcal{C}_1$ *of* Σ *in* $T^*X - 0$ *and a homogeneous canonical transformation* $f : \mathcal{C} \to \mathcal{C}_1$ *such that* f *is the identity on* Σ *and* $f_*\Lambda = \Lambda_1$.

Proof. Given a symplectic vector space V let $\mathfrak{M}(V)$ be the set of all positive definite Lagrangian subspaces of $V \otimes \mathbb{C}$. As an abstract manifold $\mathfrak{M}(V)$ is diffeomorphic to $Sp(n)/U(n)$ (see [27]); so in particular it is contractible. Let $\mathfrak{M} \to \Sigma$ be the fiber bundle over Σ whose fiber at p is $\mathfrak{M}(N_p)$, N_p being the normal space to Σ at p. Given sections $\vartriangle$ and $\vartriangle_1$ of $\mathfrak{M}$ we can find a homotopy, $\vartriangle_t$, between them because of the contractibility of the fiber. Hence if Λ and Λ_1 are positive definite Lagrangian subbundles of $N \otimes C$ we can find a family $\{\Lambda_t\}$ of positive definite Lagrangian subbundles of $N \otimes C$ depending smoothly on t such that $\Lambda = \Lambda_t$, $t = 0$, and $\Lambda_1 = \Lambda_t$, $t = 1$. Let us try to construct a homogeneous canonical transformation f_t such that f_t = identity on Σ and $(f_t)_*\Lambda = \Lambda_t$. We will first show that there exists an automorphism $A_t: N \to N$ of symplectic vector bundles mapping Λ onto Λ_t and depending smoothly on t. Let $\mathcal{G}$ be the fiber bundle over Σ whose fiber at p is the set of all symplectic linear mappings of N_p.

Consider the fiber mapping

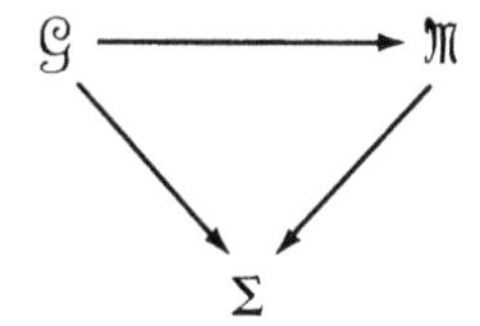

which maps $A \in \mathcal{G}_p$ onto $A(\Lambda_p)$. By the covering homotopy theorem, the section $\lambda_t : \Sigma \to \mathfrak{M}$ defined above lifts to a section $\tilde{\lambda}_t : \Sigma \to \mathcal{G}$ such that $\tilde{\lambda}_0$ is the "identity" section of $\mathcal{G}$. For each t, $\tilde{\lambda}_t$ can be viewed as a vector bundle isomorphism $A_t : N \to N$ mapping Λ onto Λ_t. This establishes the existence of A_t.

Consider $A_t^{-1}\dot{A}_t$ at $p \in \Sigma$. This is an element of the infinitesimal symplectic group, $sp(N_p)$; so, by results of §4, it determines a unique quadratic form $Q_t(p)$ on N_p. Let g_t be a function on $T^*X - 0$ which vanishes to second order Σ and whose Hessian at $p \in \Sigma$ is $Q_t(p)$. We will leave it as an easy exercise for the reader to show that such a g_t exists and can be made to depend smoothly on t. Also we will let the reader convince himself that the whole construction above can be carried out in such a way that the g_t we end up with is homogeneous of degree one.

Let Ξ_t be the Hamiltonian vector field associated with g_t. Define f_t by integrating the equation

$$f_t^{-1}(df_t/dt) = \Xi_t$$

with initial data f_o = identity. Then $f = f_1$ satisfies the hypotheses of the lemma. Q.E.D.

We will use this lemma to prove

PROPOSITION 14.6. *The Hilbert polynomial associated with* Σ, Φ *and* π *is independent of* π.

Proof. As above let there be given two Toeplitz structures on Σ and let π and π_1 be their Szegö projectors, with $\sigma(\pi) = \pi_\Lambda$ and $\sigma(\pi_1) = \pi_{\Lambda_1}$. Let f be as in Lemma 14.5 and let F be a Fourier integral operator associated with f such that $I - F^*F$ is smoothing in a conic neighborhood of Σ. Let $Q = F^*F$. Then $\pi Q \pi = \pi + K$, K being a smoothing operator. Replacing, if necessary, the range of π by a finite codimension subspace we can assume that the L^2 norm of K is < 1. With this assumption there exists a smoothing operator L such that $(I+L)^2 = (I+K)^{-1}$. Replacing F by $F(I+L)$, we have $\pi F^* F \pi = \pi$. Let $\pi_1 = F\pi F^*$. Then $\pi_1 = \pi_1^2 = \pi_1^*$; so π_1 defines a Toeplitz structure on Σ for which $\sigma(\pi_1) = \pi_{\Lambda_1}$. Moreover if T is a Toeplitz operator associated with π, FTF^* is a Toeplitz operator associated with π_1 by Lemma 14.4. Moreover it has the same symbol and the same spectrum as T. Now apply this remark to a Toeplitz operator, T, whose spectrum is contained in the union of the intervals $(13.4)_k$. Q.E.D.

To show that the Hilbert polynomial is strictly a symplectic invariant of the pair Σ, Φ, we must still show that it does not depend on how Σ is imbedded as a symplectic submanifold of a cotangent bundle.

PROPOSITION 14.7. *Let X_1 be a compact metalinear manifold (not necessarily of the same dimension as X). Let Σ_1 be a closed homogeneous symplectic submanifold of T^*X_1 and let Φ_1 be a smooth homogeneous function of degree one on Σ_1. Suppose there exists a homogeneous symplectic diffeomorphism, $f: \Sigma \to \Sigma_1$, such that $f^*\Phi_1 = \Phi$. Then the Hilbert polynomial associated with (Σ, Φ) is identical with the Hilbert polynomial associated with (Σ_1, Φ_1).*

Proof. If $\dim X = \dim X_1$ the proof of this is more or less the same as the proof of Proposition 14.6, so we will just consider the case, $\dim X < \dim X_1$. Suppose $m = \dim X_1 - \dim X$. Let Q be a positive self-adjoint pseudodifferential operator of order one on X such that $\sigma(Q) > 0$ everywhere and $[Q, \pi] = 0$. On $X \times \mathbf{R}^m$ consider the system of pseudodifferential operators

(14.2) $$D_i = (1/\sqrt{-1})(\partial/\partial y_i + y_i Q) \quad i = 1, \cdots, m .$$

Let π_e be the orthogonal projection of $L^2(X \times R^m)$ onto the space of solutions of the system of equations, $D_i f = 0$, $i = 1, \cdots, m$. Let $\{\lambda_i\}$ be the eigenvalues of Q and $\{\phi_i\}$ the corresponding eigenfunctions. Let $F_e : L^2(X) \to L^2(X \times R^m)$ be the mapping, $F_e \phi_i = \sqrt{\lambda_i/\pi}\, e^{-\lambda_i y^2/2} \phi_i$. Then $F_e^* F_e = I$ and $F_e F_e^* = \pi_e$. Let $e : T^*X - 0 \to T^*(X \times R^m) - 0$ be the mapping, $(x, \xi) \to (x, \xi, 0, 0)$. e is a symplectic imbedding of $T^*X - 0$ into $T^*(X \times R^m) - 0$. By arguments similar to those of §2 one can show:

LEMMA 14.8. F_e *is a Hermite operator associated with* e. *Moreover, if* $\Sigma_e = e(\Sigma)$, *then* $F_e \pi F_e^*$ *is the Szegö projector for a Toeplitz structure on* Σ_e.

We will omit the proof. To prove Proposition 14.7, let $\mathcal{C}$ and $\mathcal{C}_1$ be conic neighborhoods of Σ_e and Σ_1, $g : \mathcal{C} \to \mathcal{C}_1$ a homogeneous canonical transformation extending f and $F_1 : L^2(X \times R^m) \to L^2(X_1)$ a Fourier integral operator associated with g such that $F_1^* F_1 - I$ is smoothing in a conic neighborhood of Σ_e. Let $F = F_1 F_e$. Then $\pi F^* F \pi - \pi$ is smoothing. Just as in the proof of Proposition 14.6, we can modify F by post-multiplying it by an operator of the form, $I + L$, L smoothing, so that $\pi F^* F \pi = \pi$. By Lemma 14.8 $F \pi F^*$ is the Szegö projector for a Toeplitz structure on Σ_1, and it is easy to see that the Hilbert polynomial of the pair (Σ_1, Φ_1) defined by means of this Toeplitz structure is identical with the Hilbert polynomial of (Σ, Φ). Q.E.D.

We have shown above that given the pair, Σ, Φ, we can construct from the trajectories of the Hamiltonian vector field associated with Φ a compact symplectic manifold (M, ω) such that $[\omega] \in H^2(M, Z)$. We will show now that we can turn this construction around. Let (M, ω) be a symplectic manifold such that $[\omega] \in H^2(M, Z)$. For the following see [26].

LEMMA 14.9. *There exists a principal* S^1 *bundle*

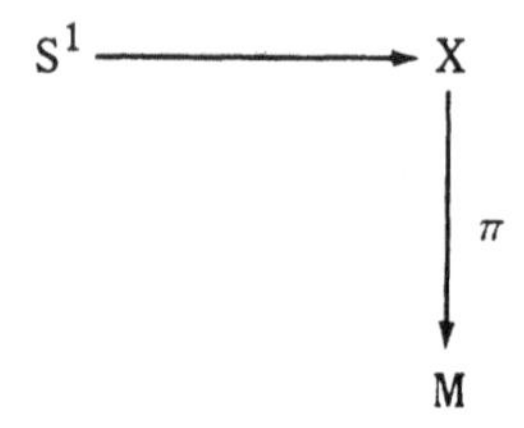

with connection α *such that* ω *is the curvature form of* α. *In addition, suppose* M *is simply connected. Then, given any two such bundles, say* (X, α) *and* (X', α') *there exists an isomorphism*

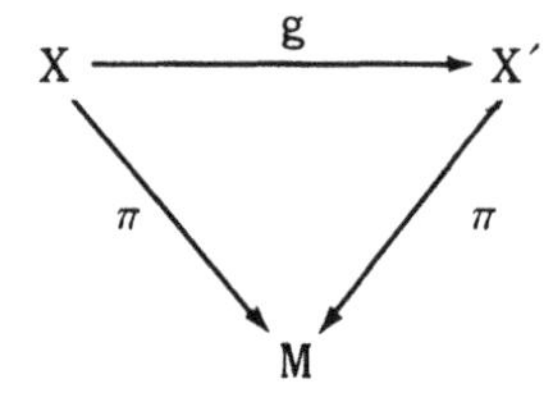

of principal S^1 *bundles such that* $g^*\alpha' = \alpha$.

Let X and α be as in this lemma. Since $d\alpha = \pi^*\omega$, $d\alpha$ is of rank $r = \dim M/2$; so in particular it is a *contact* form on X. Let Σ be the subbundle of T^*X whose fiber at $x \in X$ is all positive multiples of α_x. Σ is a homogeneous symplectic submanifold of $T^*X - 0$. Moreover, if Ξ is the infinitesimal generator of the S^1 action, then Ξ is a contact vector field on X all of whose integral curves are periodic of period 2π. The Hamiltonian function defining this contact vector field is the function $\Phi : \Sigma \to \mathbf{R}^+$ whose value at $(x, \xi) \in \Sigma$ is $\langle \Xi_x, \xi \rangle$. (Here $\Xi_x \in T_x$ and $\xi \in T^*_x$.) We will call $\Sigma \subset T^*X - 0$ and Φ the *canonical* realization of (M, ω). By Lemma 14.9 it is uniquely determined up to a homogeneous symplectic diffeomorphism.

DEFINITION. The Hilbert polynomial, p_a, associated with $\Sigma \subset T^*X - 0$, Φ, and a given Toeplitz structure on Σ, will be called the *analytic* Hilbert polynomial of the pair (M, ω).

THEOREM 14.10. *There exists an integer* k_0 *such that* $p_a(x+k_0) = p_{top}(x)$.

The proof of Theorem 14.10 will be based on a lemma whose proof will be given in the appendix. Let $\langle\ \rangle$ be an inner product on the tangent bundle of M. As we saw at the beginning of this section, we can associate with $\langle\ \rangle$ a positive definite Lagrangian subbundle, $\Lambda^{1,0}$, of $T^*M \otimes \mathbf{C}$. Let $\Lambda^{k,0}$ be the k-th alternating product of $\Lambda^{1,0}$. Given $m \in M$ and $\xi \in T^*_m$ let $\xi^{1,0}$ be the unique elements of $\Lambda^{1,0}_m$ for which $\mathrm{Re}\,\xi^{1,0} = \xi$ and let $\delta : \Lambda^{i,0}_m \to \Lambda^{i+1,0}_m$ be the map, $\mu \to \xi^{1,0} \wedge \mu$. It is clear that if $\xi \neq 0$ the sequence

$$(14.3)_\xi \qquad 0 \longrightarrow \Lambda^{0,0}_m \xrightarrow{\ \delta\ } \Lambda^{1,0}_m \longrightarrow \cdots \longrightarrow \Lambda^{r,0}_m \longrightarrow 0$$

is exact. Given $x \in X$ and $m = \pi(x)$, we have a map, $(d\pi)^t : \dot{\Lambda}^k T^*_m \to \Lambda^k T^*_x$. Let us denote by $\Lambda^{k,0}_b$ the vector subbundle of $\Lambda^k T^* X \otimes C$ whose fiber at $x \in X$ is $(d\pi)^t \dot{\Lambda}^{k,0}_m$. Given $\xi = T^*_x$ we can write $\xi = \xi_0 + \mathrm{Re}\,\xi^{1,0}$ with $\xi_0 = c\alpha_x$ and $\xi^{1,0} \in (\Lambda^{1,0}_b)_x$, and as above, we get a sequence

$$(14.4)_\xi \qquad 0 \longrightarrow (\Lambda^{0,0}_b)_x \xrightarrow{\ \delta\ } (\Lambda^{1,0}_b)_x \longrightarrow \cdots \xrightarrow{\ \delta\ } (\Lambda^{r,0}_b)_x \longrightarrow 0 .$$

which is exact providing $\xi \neq c\alpha_x$. The action of S^1 on X induces an action of S^1 on the vector bundles $\Lambda^k T^* X$. It is clear that the $\Lambda^{k,0}_b$ are preserved by this action and that this action transforms the sequence $(14.4)_\xi$ in an equivariant way.

Given a C^∞ section, f, of $\Lambda^{k,0}_b$, we can associate to each integer n another section, f_n, of $\Lambda^{k,0}_b$ by the formula

$$f_n(x) = (1/2\pi) \int f(e^{i\theta} x)\, e^{-in\theta}\, d\theta .$$

Let $C^\infty_+(\Lambda^{k,0}_b)$ be the set of all $f \in C^\infty(\Lambda^{k,0}_b)$ for which $f_n \equiv 0$ when $n < 0$. The following will be proved in §5 of the appendix.

LEMMA 14.11. *There exists a complex of first order pseudodifferential operators*

$$(14.5) \quad 0 \longrightarrow C^\infty(\Lambda_b^{0,0}) \xrightarrow{D} C^\infty(\Lambda_b^{1,0}) \longrightarrow \cdots \longrightarrow C^\infty(\Lambda_b^{r,0}) \longrightarrow 0$$

whose symbol sequence at $\xi \in T_x^* - 0$ *is the sequence* $(14.4)_\xi$. *The* D's *in this complex are equivariant with respect to the action of* S^1, *and if* π *is orthogonal projection onto the* L^2 *closure of* $\operatorname{Ker} D : C_+^\infty(\Lambda_b^{0,0}) \to C_+^\infty(\Lambda_b^{1,0})$, *then* π *is the Szegö projector for a Toeplitz structure on* Σ.

The complex (14.5) contains the subcomplex

$$(14.5)_+ \quad 0 \longrightarrow C_+^\infty(\Lambda_b^{0,0}) \longrightarrow C_+^\infty(\Lambda_b^{1,0}) \longrightarrow \cdots \longrightarrow C_+^\infty(\Lambda_b^{r,0}) \longrightarrow 0 .$$

LEMMA 14.12. *The cohomology of the complex* $(14.5)_+$ *is finite dimensional except in dimension zero.*

Proof. Since $\Lambda^{1,0}$ is a positive definite Lagrangian subbundle of $T^*M \otimes C$ the Levi form of the complex $(14.5)_+$ is positive definite; so $D + D^t$ is subelliptic of order $-\frac{1}{2}$ except in dimension zero. (See for instance [31] or [34].) Q.E.D.

Since X is a principal S^1 bundle over M we can associate with the unitary character, $\theta \to e^{i\theta}$, of S^1 a complex Hermitian line bundle L over M. Consider the space of functions, f, on X which transform under the action of S^1 according to the law

$$(14.6) \qquad f(e^{i\theta}x) = e^{ik\theta}f(x) .$$

k being a fixed integer. This space of functions can be identified in an obvious way with the space of sections of $\overset{k}{\otimes} L^*$. More generally the space of sections of $\Lambda_b^{i,0}$ which transform under the action of S^1 according to the law (14.6) can be identified with the space of sections of $\overset{k}{\otimes} L^* \otimes \Lambda^{i,0}$. Therefore, for each integer k, we get a subcomplex of (14.5) of the form

$$(14.7)_k \qquad 0 \longrightarrow C^\infty(\overset{k}{\otimes} L^*) \longrightarrow C^\infty(\overset{k}{\otimes} L^* \otimes \Lambda^{1,0}) \longrightarrow \cdots \longrightarrow C^\infty(\overset{k}{\otimes} L^* \otimes \Lambda^{r,0}) \longrightarrow 0\,.$$

LEMMA 14.13. *The complex $(14.7)_k$ is a complex of first order pseudo-differential operators on* M. *The symbol sequence of this complex at $\xi \in T^*_m$ is the sequence*

$$(14.8)_k \quad 0 \longrightarrow \overset{k}{\otimes} L^*_m \xrightarrow{\ \delta\ } \overset{k}{\otimes} L^*_m \otimes \Lambda^{1,0}_m \xrightarrow{\ \delta\ } \cdots \xrightarrow{\ \delta\ } \overset{k}{\otimes} L^*_m \otimes \Lambda^{r,0}_m \longrightarrow 0$$

the δ's *being as in (14.3).*

Proof. Let $F: C^\infty(M) \to C^\infty(X)$ be the pull-back mapping, $f \in C^\infty(M) \to \pi^* f$. F is a Fourier integral operator associated with the canonical relation

$$\Gamma = \{(m, \eta, x, \xi), m = \pi(x), \xi = (d\pi)^t \eta\}\,.$$

Similarly $F^*: C^\infty(X) \to C^\infty(M)$ is a Fourier integral operator associated with Γ^{-1}. Let Q be a pseudodifferential operator of order m on X. The "clean" composition formulas for Fourier integral operators developed in [12] (see also §9) tell us that F^*QF is a Fourier integral operator of order m on M associated with the identity canonical relation, or, in other words, a pseudodifferential operator of order m. More generally let $F_{k,i}$ be the inclusion map of $C^\infty(\overset{k}{\otimes} L^* \otimes \Lambda^{i,0})$ into $C^\infty(\Lambda_b^{i,0})$. Locally $F_{k,i}$ is the composition of a vector bundle mapping and the pull-back mapping, F; so the same argument shows that if D is a pseudodifferential operator from $\Lambda_b^{k,0}$ to $\Lambda_b^{k+1,0}$, $F^*_{k,i+1} D F_{k,i}$ is a pseudodifferential operator from $\overset{k}{\otimes} L^* \otimes \Lambda^{i,0}$ to $\overset{k}{\otimes} L^* \otimes \Lambda^{i+1,0}$. This shows that $(14.7)_k$ is a complex of first order pseudodifferential operators. To show that its symbol sequence is as indicated, let m be a point of M, U a neighborhood of m, $\vartriangle$ a section of $\overset{k}{\otimes} L^* \otimes \Lambda^{i,0}$ supported in U and f a function on U with $df \neq 0$. Compute

$$\lim_{\tau\to\infty} \tau^{-1} e^{-\sqrt{-1}\tau f} D e^{\sqrt{-1}\tau f} \Delta$$

viewing Δ as a section of $\Lambda_b^{i,0}$ and f as a function on X and using the fact that the symbol sequence of the complex (14.5) is given by (14.4).

Q.E.D.

Let Ξ be the infinitesimal generator of the circle group action on X, let Q be the differential operator, $f \in C^\infty(X) \to (1/\sqrt{-1})\Xi f$, and let T be the Toeplitz operator, $\pi Q \pi$. It is clear that $\sigma(T) = \Phi$. Moreover, since Q and π commute, $\exp \sqrt{-1}\, 2\pi T = I$; so the spectrum of T is a sequence of integers bounded below by some integer k_0.

LEMMA 14.14. *The* k*-th eigenspace of* T *is identical with the zeroth cohomology group of the complex* $(14.7)_k$.

Proof. An element of the zeroth cohomology group of $(14.7)_k$ is an element, f, of $\ker \pi$ which transforms according to (14.6). Differentiating (14.6) we get, $Qf = kf$. Q.E.D.

LEMMA 14.15. *For* k *sufficiently large the dimension of the zeroth cohomology group of* $(14.7)_k$ *is equal to its Euler number.*

Proof. For $k > 0$, $(14.7)_k$ is a subcomplex of $(14.5)_+$, so by Lemma 14.12 all the cohomology groups of $(14.7)_k$ are zero for k sufficiently large.

Q.E.D.

The following is a corollary of the Atiyah-Singer index theorem. (See [1].)

LEMMA 14.16. *The Euler number of the complex* $(14.7)_k$ *is equal to*

$$(e^{-kc(L)}\tau)[M] \; ,$$

c(L) *being the Chern class of the line bundle,* L.

Since $c(L) = [\omega]$, this concludes the proof of Theorem 14.10.

§15. SOME CONCLUDING REMARKS

1. If $\Sigma = T^*X - 0$ then the ring of Toeplitz operators associated with Σ is the usual ring of pseudodifferential operators on X. Hence the pseudodifferential counterparts of the theorems of §13 and §14 are in fact just special cases of these theorems.

2. Let Σ be a closed homogeneous submanifold of $T^*X - 0$. Σ is called *co-isotropic* if, for each $z \in \Sigma$, $(T_z\Sigma)^\perp \subset T_z\Sigma$. One can show that if Σ is co-isotropic, the subbundle of the tangent bundle of Σ whose fiber at z is $(T_z\Sigma)^\perp$ is an *integrable* subbundle, hence defines a foliation of Σ. Let us assume this foliation is in fact a fibration, i.e. assume there is a smooth fibration, $\rho : \Sigma \to Z$, Z being a Hausdorff manifold, whose fibers are the leaves of the foliation. Let Γ be the fiber product, $\{(a,b) \in \Sigma \times \Sigma, \rho(a) = \rho(b)\}$. One can show that Γ is a canonical relation in $(T^*X - 0) \times (T^*X - 0)$. In some recent work on symplectic actions of Lie groups, Alan Weinstein introduced an orthogonal projection operator, $\pi : L^2(X) \to L^2(X)$, π belonging to the ring of Fourier integral operators associated with Γ. Guillemin and Sternberg have shown that for this "Weinstein projector" the set of Fourier integral operators

$$\mathcal{R} = \{\pi Q \pi, \ Q \text{ a pseudodifferential operator}\}$$

is, in fact, a ring endowed with a symbol map

$$\mathcal{R} \xrightarrow{\sigma} C^\infty(Z) .$$

Z inherits from Σ a natural cone structure. Let $S^k(Z)$ be the set of C^∞ functions on Z which are homogeneous of degree k and let $\mathcal{R}^k$ be the set of operators $\{\pi Q \pi, \text{ order } Q \le k\}$. Then σ is an epimorphism of

$\mathcal{R}^k$ onto $S^k(Z)$, and its kernel is $\mathcal{R}^{k-1}$. One can easily show that if $k > 0$ and $T \in \mathcal{R}^k$ is self-adjoint and has a positive symbol, then T has discrete spectrum with $+\infty$ as its only point of accumulation. Guillemin and Sternberg have shown that the analogues of many of the results of §§13-14 are true for such a T (for instance the Weyl estimate on the number of eigenvalues $\leq \lambda$, the clustering result of Helton, the theorems of Weinstein and Colin de Verdiére described in §13 and so on). See [20].

3. Let D_j, $j = 1, \cdots, p$ be the operator (2.1) and define $H^{(m)}$ to be the following subspace of $L^2(\mathbf{R}^n)$:

$$H^{(m)} = \{f \in L^2(\mathbf{R}^n), D^\alpha f = 0, \forall |\alpha| \leq m\} . \tag{15.1}$$

Let $\pi^{(m)}$ be the orthogonal projection of $L^2(X)$ onto $H^{(m)}$. When $m = 1$, $\pi^{(m)}$ is the infinitesimal model for the Szegö projector, which we studied in detail in §2. Just as in §2 we can write down quite explicitly the Schwartz kernel for $\pi^{(m)}$ and verify by inspection that it belongs to the same space of Hermite operators as $\pi^{(1)}$.

More generally, given a closed homogeneous symplectic submanifold, Σ, of $T^*X - 0$ we can associate with Σ a generalized Toeplitz structure consisting of a sequence of orthogonal projection operators

$$0 = \pi_\Sigma^{(0)} < \pi_\Sigma^{(1)} < \pi_\Sigma^{(2)} < \cdots$$

satisfying the following microlocal property:

(A) For every point $p \in \Sigma$ there exists a conic neighborhood, U, of p in $T^*X - 0$, a conic open set U_0 in $T^*\mathbf{R}^n - 0$, a homogeneous canonical transformation, $\phi : U \to U_0$, and a zeroth order Fourier integral operator, $F : C^\infty(X) \to C^\infty(\mathbf{R}^n)$ associated with ϕ such that $I - F^*F$ and $F^*\pi^{(m)}F - \pi_\Sigma^{(m)}$ are smoothing in a conic neighborhood of p.

It follows automatically from (A) that $\pi_\Sigma^{(m)}$ belongs to the ring $OP\mathcal{H}^0(X, \Sigma)$. To describe its symbol we must go back to some generalities about the metaplectic representation. Let V be a symplectic vector space and $\mathcal{S}$ the space of symplectic spinors associated with V. Given $v \in V$ let $d\rho(v)$ be the action of the corresponding element of the Heisenberg algebra on $\mathcal{S}$. (See §4.) Let Λ be a positive definite Lagrangian subspace of $V \otimes \mathbb{C}$ and set

$$E_\Lambda^{(m)} = \{f \in \mathcal{S}, d\rho(v)^m f = 0, \forall \Lambda \in V\} .$$

$E_\Lambda^{(m)}$ is a finite dimensional subspace of $\mathcal{S}$. We will denote by $\pi_\Lambda^{(m)}$ the orthogonal projection of $\mathcal{S}$ onto $E_\Lambda^{(m)}$.

More generally, if Σ is a homogeneous symplectic submanifold of $T^*X - 0$ and Λ a Lagrangian subbundle of the normal bundle of Σ we will denote by $E^{(m)}$ the vector bundle over Σ whose fiber at $z \in \Sigma$ is the subspace $(E_\Lambda^{(m)})_z$. We will denote by $(\pi_\Lambda^{(m)})_z$ the orthogonal projection of the space of symplectic spinors at z onto $(E_\Lambda^{(m)})_z$. Theorem 11.2 has the following generalization

PROPOSITION 15.1. *Let Λ be the Lagrangian subbundle of the complexified normal bundle of Σ associated with the given Toeplitz structure on Σ. Then*

$$\sigma(\pi_\Sigma^{(m)}) = \pi_\Lambda^{(m)} \tag{15.2}$$

Let Q be a pseudodifferential operator on X. An operator of the form $\pi_\Sigma^{(m)} Q \pi_\Sigma^{(m)}$ will be called a *generalized Toeplitz operator of type* m. We will say that it is of order r if it is of order r as a Hermite operator. Its symbol is a section of the vector bundle $\mathrm{Hom}(E^{(m)}, E^{(m)})$, and one can show that it behaves very much like an ordinary pseudodifferential operator on vector bundles.

One very interesting example of such an operator is the following. Let Q be a self-adjoint second order pseudodifferential operator on X whose symbol, q, is everywhere non-negative and is positive on $T^*X - 0$ except on Σ where it vanishes exactly to second order. This means that the Hessian of q at $z \epsilon \Sigma$ is a non-degenerate positive definite quadratic form on $N_z\Sigma$. From the symplectic structure on $N_z\Sigma$ and from this quadratic form we get a canonically associated positive definite Lagrangian subspace, Λ_z, of $N_z\Sigma \otimes \mathbb{C}$. (See §14.) By Lemma 14.5 we can arrange that this is the same as the Lagrangian space Λ_z in (15.2). Associated with the Hessian of q we have an infinitesimal linear symplectic mapping $(A_q)_z : N_z\Sigma \to N_z\Sigma$ which preserves Λ_z and hence induces a map

$$(A_q^{(m)})_z : E_z^{(m)} \to E_z^{(m)} .$$

PROPOSITION 15.2. *The operator* $T^{(m)} = \pi_\Sigma^{(m)} Q \pi_\Sigma^{(m)}$ *is a first order Toeplitz operator of type* m, *and its symbol is*

$$(1/\sqrt{-1})A_q^{(m)} + \sigma_{\text{sub}}(Q) I . \tag{15.3}$$

Proof. The formula (15.9) is a special case of the formula (10.3). Q.E.D.

For each $z \epsilon \Sigma$ the eigenvalues of $(A_q^{(m)})_z$ are easy to compute using elementary facts about the metaplectic representation. As we remarked above $(A_q)_z$ maps Λ_z into Λ_z, and because of the positivity of the Hessian of q at z the eigenvalues of this map are of the form $\sqrt{-1}\lambda_i$, $i = 1, \cdots, p$, $\lambda_i > 0$. (Here $2p = \text{codim } \Sigma$.) One can prove

PROPOSITION 15.3. *The eigenvalues of* $A_q^{(m)} : E^{(m)} \to E^{(m)}$ *at* z *are the numbers*

$$\sum_{i=1}^{p} a_i(\lambda_i + 1/2) \tag{15.4}$$

the a_i*'s being non-negative integers with* $a_1 + \cdots + a_p \leq m$.

COROLLARY. *If* $\sigma_{\text{sub}}(Q) + \frac{1}{2}\sum_{i=1}^{p} \lambda_i$ *is everywhere positive then* $T^{(m)}$ *has discrete spectrum with only* $+\infty$ *as a point of accumulation.*

Proof. If the expression above is everywhere positive, the symbol of $T^{(m)}$ is everywhere positive definite by Proposition 15.2; so the spectrum of $T^{(m)}$ is bounded from below and $T^{(m)}$ admits a compact parametrix.

Q.E.D.

Let $N^{(m)}(\lambda)$ be the number of eigenvalues of $T^{(m)}$ which are $\leq \lambda$. By arguments similar to those of §13 one can prove:

THEOREM 15.4. *If the eigenvalues of* $A_q : \Lambda_z \to \Lambda_z$ *occur with constant multiplicity (not depending on* z*) then*

$$N^{(m)}(\lambda) = \left(\frac{1}{2\pi}\right)^r \sum_{|\alpha| \leq m} \text{Volume}(\Sigma_\alpha)\lambda^r + O(\lambda^{r-1}), \tag{15.5}$$

$2r$ *being the dimension of* Σ *and* Σ_α *the submanifold of* Σ *on which*

$\sigma_{\text{sub}}(Q) + \sum_{i=1}^{p} \alpha_i \lambda_i \leq 1$.

From the results above, one can obtain a certain amount of information about the spectrum of Q itself. It is known that if $\sigma_{\text{sub}}(Q) + \frac{1}{2}\sum \lambda_i > 0$, Q is hypoelliptic of order one; so it has discrete spectrum with $+\infty$ as the only point of accumulation. Let $N(\lambda)$ be the number of eigenvalues of Q less than or equal to λ. By construction, Q differs from the operator, $\pi^{(m)} Q \pi^{(m)} + (1-\pi^{(m)}) Q (1-\pi^{(m)})$, by a bounded operator. If C is the L^2 norm of this operator then by standard mini-max arguments, (see [11]),

$$N(\lambda) \geq N^{(m)}(\lambda - C) = N^{(m)}(\lambda) + O(\lambda^{r-1}) \tag{15.6}$$

the equality on the right being a consequence of (15.5); so (15.5) gives an asymptotic lower bound on $N(\lambda)$. Menikoff and Sjöstrand [29] have recently obtained some very general estimates on the asymptotic behavior of eigenvalues of hypoelliptic operators which show that the asymptotic lower bound (15.6) is in fact an asymptotic upper bound as well when $\dim \Sigma \geq \dim X$. It would perhaps be illuminating to find a proof of their result, in the special case considered here, using Toeplitz operators.

BIBLIOGRAPHY

[1] M. F. Atiyah and I. M. Singer, "The index of elliptic operators I," Ann. Math. 87 (1968), 484-530.

[2] R. Blattner, "Quantization and representation theory," Proceedings, AMS Symposium in Pure Math., Vol. 26 (1974).

[3] R. Bott, "On the iteration of closed geodesics and the Sturm intersection theory," Comm. Pure Appl. Math., 9 (1956), 176-206.

[4] L. Boutet de Monvel, "Hypoelliptic operators with double characteristics and related pseudodifferential operators," Comm. Pure Appl. Math., 27 (1974), 585-639.

[5] ———, "On the index of Toeplitz operators of several complex variables, Invent. Math., 50 (1979), 249-272.

[6] ———, "Nombre de valeurs propres d'un operateur elliptique et polynome de Hilbert-Samuel," Seminare Bourbaki, 1978-79, no. 532.

[7] L. Boutet de Monvel and J. Sjöstrand, "Sur la singularité des noyaux de Bergmann et de Szegö," Asterisque 34-35 (1976), 123-164.

[8] J. Chazarain, "Formule de Poisson pour les variétés riemanniennes," Invent. Math., 24 (1974), 65-82.

[9] Y. Colin de Verdiére, "Spectre du laplacien et longueurs des geodesiques periodiques," Comp. Math., 27 (1974), 159-184.

[10] ———, "Sur le spectre des operateurs elliptiques a bicaracteristiques toutes periodiques," Commentarii Math., Helvetici, 54 (1979), 508-522.

[11] R. Courant and D. Hilbert, *Methods of Mathematical Physics* Volume II, Interscience, New York, (1962).

[12] J. J. Duistermaat and V. Guillemin, "The spectrum of positive elliptic operators and periodic geodesics," Invent. Math., 29 (1975), 184-269.

[13] J. J. Duistermaat and L. Hörmander, "Fourier Integral Operators II," Acta Math., Vol. 128 (1972), 183-269.

[14] Yu. V. Egorov, "On canonical transformations of pseudodifferential operators," Uspehi Mat. Nauk., 25 (1969), 235-236.

[15] V. Guillemin, "Symplectic spinors and partial differential equations," Coll. Inst. CNRS no. 237 Géométrie symplectique et physique mathematique, 217-252.

[16] ———, "Lectures on spectral theory of elliptic operators," Duke Math. Journal, vol. 44, no. 3 (1977), 485-517.

[17] ———, "Some classical theorems in spectral theory revisited," in *Seminar on singularities of solutions of linear partial differential equations*, Annals of Math. Studies no. 91, Princeton University Press, Princeton, N. J. (1979), 219-259.

[18] V. Guillemin and S. Sternberg, *Geometric Asymptotics*, AMS Math. Surveys, no. 14, Providence, R. I. (1978).

[19] ———, "On the spectra of commuting pseudodifferential operators," in *Partial Differential Equations and Geometry Proceedings of the Park City Conference*, Marcel Dekker, Inc., New York, N. Y. (1979), 149-164.

[20] ———, "Some problems in integral geometry and some related problems in micro-local analysis," Amer. Journal of Math., Vol. 101 (1979), 915-955.

[21] W. Helton, "An operator approach to partial differential equations, propagation of singularities and spectral theory," Indiana University Math., Journal Vol. 26, no. 6 (1977), 997-1018.

[22] F. Hirzebruch, *Neue Topologische Methoden in der Algebraischen Geometrie*, Springer Verlag, Berlin (1956).

[23] L. Hörmander, "Pseudodifferential operators and non-elliptic boundary problems," Ann. of Math., Vol. 83, no. 1 (1966), 129-209.

[24] ———, "The spectral function of an elliptic operator," Acta Math., 121 (1968), 193-218.

[25] ———, "Fourier integral operators I," Acta Math., 127 (1971), 79-183.

[26] B. Kostant, "Quantization and unitary representations," in *Lectures in Modern Analysis and Applications*, Springer Lecture Notes, no. 170 (1970), 87-208.

[27] ———, "Symplectic Spinors," Conv. di Geom. Simp. e Fis. Math., INDAM Rome (1973).

[28] A. Melin and J. Sjöstrand, "Fourier integral operators with complex valued phase functions," in *Fourier Integral Operators and Partial Differential Equations*, Lectures Notes no. 459, Springer Verlag, 120-223.

[29] A. Menikoff and J. Sjöstrand, "On the eigenvalues of a class of hypoelliptic operators," Math. Ann., Vol. 235 (1978), 55-85.

[30] D. Mumford, *Algebraic Varieties I, Complex Projective Varieties*, Springer-Verlag, New York (1976).

[31] C. Rockland, "Poisson complexes and subellipticity," J. Diff. Geom., Vol. 9, no. 1 (1974), 71-91.

[32] I. E. Segal, "Transforms for operators and symplectic automorphisms over a locally compact abelian group," Math. Scand. 13 (1963), 31-43.

[33] N. Steenrod, *The Topology of Fiber Bundles*, Princeton University Press, Princeton, N. J. (1951).

[34] W. Sweeney, "A condition for subellipticity in Spencer's Neumann problem," Journal of Diff. Equations, Vol. 21, no. 2 (1976), 316-362.

[35] A. Weinstein, "Asymptotics of eigenvalue clusters for the Laplacian plus a potential," Duke Math. J., Vol. 44 (1977), 883-892.

[36] A. Weil, Variétés Kählériennes, Hermann, Paris (1958).

[37] H. Widom, "Eigenvalue distribution theorems in certain homogeneous spaces," J. Funct. Anal. (to appear).

APPENDIX: QUANTIZED CONTACT STRUCTURES

A.0. In the preceding chapters, we have used generalized Szegö projectors and generalized Toeplitz operators associated with arbitrary symplectic subcones of the cotangent bundle of a manifold, and occasionally a complex of first order pseudodifferential operators playing the same role as the $\bar{\partial}_b$ complex for the Szegö projector. The purpose of this appendix is to show how these are constructed. We will use the Fourier integral operators of A. Melin and J. Sjöstrand [7], although for most purposes the Hermite operators and their symbolic calculus are enough, and their use leads to somewhat simpler and shorter proofs; but some of the finer properties (cf. A.5) would no longer hold for more general Hermite operators.

We will say that a symbol $p(x, \xi)$ of degree m is regular if it admits an asymptotic expansion:

$$p \sim \sum_{j=0}^{\infty} p_{m-j}(x, \xi)$$

where p_{m-j} is homogeneous of degree $m-j$ with respect to ξ, j a positive integer; the degree may be any number, although $4m$ will usually be an integer, and in the case of pseudodifferential operators, m will be an integer; but the degrees go down by integral steps.

Unless otherwise specified, pseudodifferential operator means "regular" (or classical) ps.d.o., i.e. the symbol is regular, and the degree is an integer.

If P is a pseudodifferential operator with total symbol $p(x,\xi) \sim \sum p_{m-j}(x,\xi)$, we denote by $\tilde{P}$ any pseudodifferential operator with total symbol

$$p(x,\xi) \sim \sum (-1)^{m-j}\, p_{m-j}(x,-\xi)$$

this definition does not depend on the choice of local coordinates and still makes sense for pseudodifferential operators on sections of smooth vector bundles. For example we have $P = \tilde{P}$ if P is a differential operator or the parametrix of an elliptic differential operator. One checks easily $\widetilde{PQ} = \tilde{P}\tilde{Q}$.

A pseudodifferential operator of type $\frac{1}{2}$ is a (nonclassical) pseudodifferential operator whose total symbol satisfies locally

$$|(\partial/\partial x)^{\alpha}(\partial/\partial\xi)\, p(x,\xi)| \le C_{\alpha\beta}(1+|\xi|)^{m+½|\alpha|-½|\beta|}$$

for suitable constants $C_{\alpha\beta}$ (m being the degree of P).

A.1. *Contact manifolds and symplectic cones*

A symplectic cone Σ is a cone (i.e. a C^{∞} paracompact manifold with a free action of the multiplicative group $\mathbf{R}_+^{\times}$ of positive real numbers) equipped with a symplectic 2-form ω (i.e. ω is non-degenerate and $d\omega = 0$) which is homogeneous of degree 1.

Such a cone is isomorphic, as a C^{∞} manifold, to $X \times \mathbf{R}_+^{\times}$ with the standard action of $\mathbf{R}_+^{\times}$ ($\lambda \cdot (x,r) = (x,\lambda r)$). If r denotes the 2nd projection (r is a positive C^{∞} function), ω can be written in a unique manner:

$$\omega = r\,\omega_2 + dr \wedge \omega_1$$

where ω_1 (resp. ω_2) is the pull-back of a 1-form (resp. a 2-form) on X. The equality $d\omega = 0$ implies $\omega = d(r\,\omega_1)$. The 1-form

$$\lambda = r\omega_1$$

is the unique 1-form which is homogeneous of degree 1, orthogonal to the radial vector field $\partial/\partial r$, and such that $\omega = d\lambda$. We will call it the Liouville form of Σ.

An important example of a symplectic cone is the cone $T^*X - 0$ (cotangent bundle minus the zero section), where X is a C^∞ manifold, equipped with its canonical symplectic structure: in local coordinates

$$\omega = \Sigma d\xi_j \wedge dx_j, \quad \lambda = \Sigma \xi_j \wedge dx_j$$

where the x_j's are a system of local coordinates on X, the ξ_j's the corresponding linear coordinates on the fibers of T^*X.

An oriented contact manifold X is a C^∞ (paracompact) manifold of odd dimension $2n-1$, equipped with a halfline subbundle $\Sigma \subset T^*X - 0$ such that Σ is symplectic as a submanifold of $T^*X - 0$. The structure defined by a halfline subbundle of $T^*X - 0$ is equivalent to the structure defined by a class of 1-forms (the sections of Σ), two forms α, α' being equivalent if $\alpha = \phi\alpha'$ with $\phi \in C^\infty(X)$, $\phi > 0$. The cone generated by a 1-form α is symplectic if and only if α is a contact form, i.e. the $2n-1$-form $d\alpha_\wedge(\alpha)^{n-1}$ vanishes nowhere in X.

If X is an oriented contact manifold, it is isomorphic to the base of the symplectic cone which defines the contact structure. Conversely if Σ is a symplectic cone, X its base, λ its Liouville form, the 1-forms $\alpha = \psi^*(\lambda)$ with ψ a section of Σ define an oriented contact structure on X, and there is a unique isomorphism from Σ to the halfline bundle $\Sigma' \subset T^*X - 0$ defining the contact structure which preserves the projections $\Sigma \to X$ and $\Sigma' \to X$ (there is a category equivalence between symplectic cones and oriented contact manifolds).

An important example for us is the following: let W be a complex manifold, $\Omega \subset W$ an open subset with smooth boundary $X = \partial\Omega$. Ω is defined by an inequation $\rho < 0$, with $\rho \in C^\infty(W)$, $d\rho \neq 0$ on X. The 1-form $\alpha = \frac{1}{i}\,\partial\rho|_X$ is real (because $\partial\rho|_X$ and $\overline{\partial}\rho|_X$ are complex conjugate and their sum $\partial\rho|_X + \bar{\partial}\rho|_X = d\rho|_X$ vanishes so both are pure imaginary). The line bundle spanned by α is the set of all (real) covectors $\xi \in T^*X$ orthogonal to holomorphic (or antiholomorphic) vectors tangent to X. The halfline bundle $\Sigma^+ \subset T^*X - 0$ spanned by α is symplectic if and only if the Leviform of X is nondegenerate. Thus in particular the (smooth) boundary of a strictly pseudoconvex domain $\Omega \subset W$ is canonically equipped with an oriented contact structure (the structure defined by the contact form $\alpha = \frac{1}{i}\,\partial\rho|_X$).

A.2. *Adapted Fourier integral operators*

In what follows we will be dealing with Fourier integral operators associated to strictly positive $(>> 0)$ complex canonical relations. These are described in [7], §3, and we will just recall here some basic facts about them:

(2.1) Let Y be a C^∞ real manifold. An almost complex submanifold Λ of Y, of codimension n, is locally defined by n smooth complex equations $u_1 = \cdots u_n = 0$, with the differentials du_j linearly independent; the ideal $I_\Lambda \subset C^\infty(Y, \mathbf{C})$ of functions vanishing on Λ is locally generated by the u_j. The submanifold Λ is completely determined by its ideal I_Λ; we refer to [7] for its geometric interpretation as a submanifold of a complexification of Y; here Λ will also be completely determined by its Taylor expansion (jet of infinite order) along the set of real points Λ_R (defined by the equations $\mathrm{Re}\, u_j = \mathrm{Im}\, u_j = 0$) and we can manipulate Λ as if it were analytic.

(2.2) Let now Y be symplectic, of dimension $2n$. Then Λ is Lagrangian if the Poisson brackets $\{u_p, u_q\}$ vanish on Λ, i.e. belong to the ideal I_Λ.

(2.3) We will say that Λ is strictly positive $(>> 0)$ if it is Lagrangian, if the set of real points Λ_R is a smooth submanifold of codimension $n+k$ $(0 \le k \le n)$, and the hermitian matrix $\frac{1}{i}\{u_p, \bar{u}_q\}$ is positive of rank k

on Λ_R; this implies that the system of differentials du_p, $d\bar{u}_p$ is of linear rank $n+k$ exactly on Λ_R; it does not depend on the choice of the u_j.

This being so, near any point of Λ_R there exist local canonical coordinates x_j, ξ_j and a neighborhood in which Λ is defined by the equations

$$\xi_j - ix_j\xi_n = 0\,(1 \le j \le k),\ \xi_j = 0\,(k<j<n),\ x_n + \frac{i}{2}\sum_1^k x_j^2 = 0\ \text{(cf. [6])},\ \xi_n > 0 .$$

If Y is a symplectic cone and Λ homogeneous, one may choose the x_j, ξ_j homogeneous (i.e. the x_j of degree 0, the ξ_j of degree 1). In this setting, if we interpret Y as a cotangent bundle, the x_j being the base coordinates and the ξ_j the cotangent coordinates, Λ is one-half of the conormal bundle of the hypersurface $x_n + \frac{i}{2}\sum_1^k x_j^2 = 0$, corresponding to the phase function $\phi(x,\theta) = \theta\left(x_n + \frac{1}{2}i\sum_1^k x_j^2\right)$ $(\theta \in R_+)$.

(2.4) Let X be a smooth manifold of dimension n, $\phi(x,\theta)$ a non-degenerate complex phase function on $X \times R^N$ (i.e. $d\phi$ does not vanish, the differentials $d\left(\frac{\partial\phi}{\partial\theta_j}\right)$ are linearly independent, and ϕ is homogeneous of degree 1 with respect to θ). If ϕ is real, the Lagrangian cone $\Lambda \subset T^*X - 0$ it defines is the image of the critical set $C\left(\frac{\partial\phi}{\partial\theta_j} = 0,\ j=1,\cdots,N\right)$ under the differential map $(x,\theta) \mapsto (x, d_x\phi)$. If ϕ is complex, Λ is still defined as an almost complex manifold if $\operatorname{Im}\phi \ge 0$.

(2.5) We will say that ϕ is $\gg 0$ if the set of real critical points C_R (defined by the equations $\operatorname{Re}\frac{\partial\phi}{\partial\theta_j} = \operatorname{Im}\frac{\partial\phi}{\partial\theta_j} = 0$) is a smooth manifold of codimension $N+k$, and the imaginary part $\operatorname{Im}\phi$ is >0 outside of C_R, and transversally elliptic along C_R.

A Lagrangian almost complex subcone $\Lambda \subset T^*X - 0$ is $\gg 0$ if and only if it can locally be defined by a $\gg 0$ phase (cf. [7] or footnote).

(2.6) Let X, Y be two smooth real manifolds, and $\mathcal{C}$ an almost complex canonical relation from X to Y. We will say that $\mathcal{C}$ is $>> 0$ if the corresponding Lagrangian cone $\Lambda \subset T^*(X \times Y) - 0$ (image of $\mathcal{C}$ under the symmetry $(x, y, \xi, \eta) \mapsto (x, y, \xi, -\eta)$) is $>> 0$.

DEFINITION 2.7. *Let X, X' be two smooth real manifolds, $\Sigma \subset T^*X$, $\Sigma' \subset T^*X'$ two smooth symplectic subcones, and $\chi : \Sigma \to \Sigma'$ an isomorphism of symplectic cones. A Fourier integral operator from X to X' is adapted to χ if its canonical relation is $>> 0$, with real part the graph of χ, and its symbol is elliptic.*

PROPOSITION 2.8. *Let X, X', X'' be three manifolds, Σ, Σ', Σ'' three symplectic subcones of their respective cotangent bundles, $\chi : \Sigma \to \Sigma'$ and $\chi' : \Sigma' \to \Sigma''$ isomorphisms of symplectic cones, A and A' Fourier integral operators adapted to χ and χ'. Then $A' \circ A$ is adapted to $\chi' \circ \chi$ and A^* is adapted to χ^{-1}.*

The proof is immediate: let $\mathcal{C}$ and $\mathcal{C}'$ be the canonical relations corresponding to A and A'; the transversality condition ensuring that $\mathcal{C}' \circ \mathcal{C}$ is a (smooth) canonical relation (cf. [7]) and the fact that $\mathcal{C}' \circ \mathcal{C}$ is $>> 0$ essentially follow from the fact that the sum of two $>>0$ hermitian forms is still $>> 0$; the real part of $\mathcal{C}' \circ \mathcal{C}$ is the graph of $\chi' \circ \chi$ (composition of the real parts of $\mathcal{C}'$ and $\mathcal{C}$); it follows that $A' \circ A$ is a Fourier integral operator associated with $\mathcal{C}' \circ \mathcal{C}$; and on the real part of $\mathcal{C}' \circ \mathcal{C}$ the symbol of $A' \circ A$ is elliptic because it is the product of the two elliptic symbols of A and A'. The assertion for A^* is obvious and left to the reader.

PROPOSITION 2.9. *Let X, X', Σ, Σ', χ be as in definition 2.7. There exists a Fourier integral operator adapted to χ.*

Proof. Let us first suppose that X' is the basis of Σ, $\Sigma' \subset T^*X'$ the symplectic cone corresponding to its contact structure, $\chi : \Sigma \to \Sigma'$ the canonical isomorphism. We may suppose that X is embedded in $\mathbf{R}^N$, so that X' is embedded in the cotangent sphere of $\mathbf{R}^N$. Let ϕ be the phase function on $\mathbf{R}_+ \times X \times X'$, restriction of the function $t(\langle x'-x, \xi\rangle + i|x'-x|^2)$ on $\mathbf{R}_+ \times \mathbf{R}^N \times T^*\mathbf{R}^N$ (where $x \in \mathbf{R}^N$, $(x', \xi) \in T^*\mathbf{R}^N$): then ϕ is $>> 0$, and the real part of the canonical relation it defines is precisely χ. Choosing any elliptic symbol on $\mathbf{R}_+ \times X \times X'$, for instance the symbol 1, we get a Fourier integral operator adapted to χ. Taking adjoints, we get another one adapted to χ^{-1}.

In the general case, let Y be the basis of Σ, and identify Y with the basis of Σ' by means of χ; let $\Sigma'' \subset T^*Y$ be the symplectic cone corresponding to the contact structure of Y. Then we have $\chi = \chi'' \circ \chi'$ where χ' is the canonical isomorphism $\Sigma \to \Sigma''$ and χ'' the inverse of the canonical isomorphism $\Sigma' \to \Sigma''$. We have just seen that there exist Fourier integral operators A' and A'' adapted to χ' and χ'', so $A'' \circ A'$ is adapted to $\chi'' \circ \chi'$ by Proposition 2.8.

(2.10) FOOTNOTE. We give here a short proof of (2.5): first let Λ be a smooth $>> 0$ Lagrangian subcone of T^*X, and (x, ξ) a real point of Λ. We may choose local coordinates $x_1, \cdots, x_n$ near x such that Λ is defined by the equations $u_j = x_j + \dfrac{\partial U}{\partial \xi_j} = 0$, where $U(\xi)$ is some smooth function of ξ alone. Since Λ is $>> 0$, the hermitian matrix $\frac{1}{i}\{u_p, \overline{u}_q\} = 2 \operatorname{Im} \dfrac{\partial^2 U}{\partial \xi_p \partial \xi_q}$ is positive of rank k (codim $\Lambda_R = n+k$). It follows immediately that $\operatorname{Im} U$ vanishes on a smooth manifold S of codimension k, and that near S it is ≥ 0 and transversally elliptic. Now if we set $\phi(x, \xi) = x \cdot \xi + U(\xi) + i\varepsilon \Sigma u_j^2$, the Lagrangian cone defined by ϕ is precisely Λ, and ϕ is $>> 0$ (near $X \times S$) if ε is > 0 and small enough.

Conversely let $\phi(x, \theta)$ be a $>> 0$ phase function on $X \times \mathbf{R}^N$. We may again choose local coordinates $x_1, \cdots, x_n$ on X so that the $\xi_j = \dfrac{\partial \phi}{\partial x_j}$, $\mu_j = \dfrac{\partial \phi}{\partial \theta_j}$ form a local system of complex coordinates on $X \times \mathbf{R}^N$, i.e. the matrix

$$A = \begin{pmatrix} \frac{\partial \xi_j}{\partial x_k} & \frac{\partial \xi_j}{\partial \theta_k} \\ \\ \frac{\partial \mu_j}{\partial x_k} & \frac{\partial \mu_j}{\partial \theta_k} \end{pmatrix} = \begin{pmatrix} \frac{\partial^2 \phi}{\partial x_j \partial x_k} & \frac{\partial^2 \phi}{\partial x_j \partial \theta_k} \\ \\ \frac{\partial^2 \phi}{\partial x_k \partial \theta_j} & \frac{\partial^2 \phi}{\partial \theta_j \partial \theta_k} \end{pmatrix}$$

is invertible. Since ϕ is $>> 0$, Im A is positive, of rank $N+k =$ codim C_R at any point of the manifold C_R of real critical points of ϕ (with respect to θ). It follows that the same is true of $-A^{-1}$, so that the matrix $-\left(\frac{\partial x_p}{\partial \xi_q}\right)$, which is a square $n \times n$ matrix extracted from $-A^{-1}$ has its imaginary part positive, of rank $\geq k$. Now the critical (almost complex) manifold C is defined by the equations $\mu_1 = \cdots = \mu_N = 0$, so the ξ_j form a local system of complex coordinates on C, and there exist smooth functions $U_j(\xi)$ such that the $u_j = x_j + U_j(\xi)$ vanish on C. Then Λ is also defined by the equations $u_j(x, \xi) = 0$, and since it is Lagrangian, the functions $\frac{\partial U_j}{\partial \xi_k} - \frac{\partial U_k}{\partial \xi_j}$ vanish on Λ (or on C); so we can further replace U_j by $\frac{\partial U}{\partial \xi_j}$, with $U = \Sigma \xi_j U_j$. Now we have $\frac{1}{i}\{u_p, \overline{u}_q\} =$ $2 \operatorname{Im}\left(\frac{\partial^2 U}{\partial \xi_p \partial \xi_q}\right) = 2 \operatorname{Im}\left(\frac{\partial u_p}{\partial \xi_q}\right) = -2 \operatorname{Im}\left(\frac{\partial x_p}{\partial \xi_q}\right)$, and this is, as we have noticed above, positive of rank $\geq k$, so it follows that Λ is $>> 0$.

(2.11) Let A be a Fourier integral operator adapted to χ as in definition 2.7. Then using local representations of A by oscillating integrals with $>> 0$ phase functions, and the material of §3, §4, and §10, we see that A is a Hermite operator whose symbol at any point of the graph of χ can be interpreted as a linear operator between Schwartz spaces of rapidly decreasing functions of the form

$$f(v) \mapsto g(u) = \lambda \int e^{-q(u,v)} f(v)\, dv$$

with λ a constant and $q(u, v)$ a quadratic form with $>> 0$ real part.

In particular if $X = X'$, $\Sigma = \Sigma'$, $\chi = \mathrm{Id}_\Sigma$, and the symbol $\sigma(A)$ is a projector, it is necessarily of rank 1 (q is necessarily of the form $q_1(u) + q_2(v)$).

A.3. *Toeplitz structures*

Let X be a smooth compact manifold, $\Sigma \subset T^*X - 0$ a smooth symplectic subcone, of dimension $2n$. As usual $L^2(X)$ denotes the Hilbert space of half densities on X, and we will eventually identify it with a space of functions on X.

DEFINITION 3.1. *A Toeplitz structure associated with* Σ *is a closed subspace* $\mathcal{O}^0 \subset L^2(X)$ *such that the orthogonal projector* $S: L^2(X) \to \mathcal{O}^0$ *is a Fourier integral operator adapted to* Id_Σ.

Let us make the following remarks:

(3.2) The projector S is necessarily of degree 0, so it is continuous on the Sobolev space $H^s(X)$ for any real number s. We will denote by $\mathcal{O}^s$ the image $S(H^s(X))$, and $\mathcal{O}^\infty = \cap \mathcal{O}^s = \mathcal{O}^0 \cap C^\infty$.

(3.3) S is usually not a local operator. However since the graph of Id_Σ is contained in the graph of Id_{T^*X}, S is microlocal, i.e. it diminishes microsupports. If we neglect smooth functions and go to microfunctions, S defines a subsheaf of the sheaf of microfunctions, which we will usually denote by $\mathbf{\mathcal{O}}$.

(3.4) *Microlocal structure.* Here we will neglect smooth functions and work with microfunctions. Fourier integral operators need only be defined locally on the cotangent bundle.

Locally there exists an isomorphism of symplectic cones χ from T^*R^n to Σ (χ is an isomorphism of an open subcone of T^*R^n onto an open conic neighborhood in Σ of an arbitrarily given point of Σ). Let B be a Fourier integral operator adapted to χ, and $A \sim S_0 B$: A is also adapted to χ and we have $S_0 A \sim A$ (where $\sim$ means that the two operators coincide on microfunctions). Then $A^*{}_{\circ}A$ is adapted to $\mathrm{Id}_{T^*\mathbb{R}^n}$, i.e. it is an elliptic (positive) pseudodifferential operator. Now let us set

$$H \sim A \circ (A^*A)^{-1/2} .$$

Then we have

$$H^* \circ H \sim \mathrm{Id}, \quad H \circ H^* \sim S .$$

The first equality follows immediately from the definition of H and implies $HH^* \sim (HH^*)^* \sim (HH^*)^2$. For the second we first notice that we have $S \circ A \sim A$, hence $S \circ H \sim H$ and $S \circ HH^* \sim HH^*$. Now S and HH^* are both adapted to Id_Σ, of degree 0. So they are also Hermite operators with respect to Σ, whose symbols are projectors hence of rank 1 (cf. (2.11)). Since S and HH^* are both orthogonal and $S \circ HH^* \sim HH^*$, their symbols are equal, and $\sigma \sim S - HH^*$ is an orthogonal projector of degree $\leq -\frac{1}{2}$ as a Hermite operator. Finally we have $\sigma \sim \sigma^m$ for any integer m, so σ is of degree $-\infty$ and $S \sim HH^*$.

(3.5) Let X, Σ, $\mathcal{C}$ be as in definition 3.1. Let Λ be a strictly positive pseudodifferential operator on X, and let H_Λ be the Hilbert space $\Lambda^{-1/2}(L^2(X))$, with scalar product $(f|g)_\Lambda = (\Lambda f|g)$ (in fact we have $H_\Lambda = H^s(X)$ with $s = \frac{1}{2} \deg \Lambda$, but we have specified the norm). Let S_Λ be the orthonormal projector on $\mathcal{C}^s = S(H_\Lambda)$ in H_Λ. Then again S_Λ is a Fourier integral operator adapted to Id_Σ. Indeed this is a microlocal property, so let us take again the notations of (3.4) and set $\Lambda' \sim H^*\Lambda H$: this is a self-adjoint positive elliptic pseudodifferential operator on $\mathbb{R}^n$, and we have (microlocally) $S_\Lambda \sim H\Lambda'^{-1}H^*\Lambda$, and our assertion follows. (In the discussion above, s may be any real number but we require Λ to be regular.) It follows that the fact that the orthogonal projector on $\mathcal{C}^o$ is a Fourier integral operator adapted to Id_Σ does not depend on the choice of the L^2 norm; in particular for a subspace of functions it does not depend on the identification with a subspace of half-densities (i.e. on the choice of one > 0 density). One may further extend the definition to a subspace of sections of a smooth complex vector bundle.

EXERCISE. Let $\mathcal{O}^o$ be a subspace of $L^2(X)$ such that there exists a projector S' (not necessarily orthogonal) onto $\mathcal{O}^o$ which is a Fourier integral operator adapted to Id_Σ. Then $\mathcal{O}^o$ defines a Toeplitz structure associated with Σ (i.e. the orthogonal projector onto $\mathcal{O}^o$ is also adapted to Id_Σ).

(3.6) Let X, Σ, $\mathcal{O}$ be as in definition 3.1. A Toeplitz operator of degree m is an operator $\mathcal{O}^s \to \mathcal{O}^{s-m}$ of the form $u \mapsto T_Q u = S(Qu)$ where Q is a pseudodifferential operator of degree m on X. The Toeplitz operators form an algebra (the Toeplitz algebra) which is microlocally isomorphic to the algebra of pseudodifferential operators of n real variables and gives rise to a symbolic calculus which lives on Σ. This follows immediately from the microlocal analysis of (3.4), with the same arguments as in [2]: microlocally the map $T_Q \mapsto H^*QH$ is an isomorphism of the Toeplitz algebra on the algebra of pseudodifferential operators on R^n. We have

$$\sigma(T_Q) = \sigma(Q)|_\Sigma .$$

$\sigma_m(T_Q) = 0$ means that T_Q is really of degree $\leq m-1$, i.e. $T_Q = T_{Q'}$ with Q' of degree $\leq m-1$

$$\sigma(T_Q \circ T_{Q'}) = \sigma(T_Q)\sigma(T_{Q'})$$

$$\sigma([T_Q, T_{Q'}]) = -i\{\sigma(T_Q), \sigma(T_{Q'})\}_\Sigma$$

where $\{\ \}_\Sigma$ is the Poisson bracket on Σ.

A Toeplitz operator T_Q of degree m is elliptic if its symbol is invertible. If this is so T_Q has a parametrix $T_{Q'}$ which is a Toeplitz operator of degree $-m$, i.e. we have $T_Q T_{Q'} \sim T_{Q'} T_Q \sim Id$, $\sim$ meaning that the difference is of degree $-\infty$.

Similarly one defines an elliptic system (matrix) of Toeplitz operators: the symbol is an invertible square matrix. Such a system has an index, which is again given by the formula of Theorem 1 of [2] and is proved in the same manner (cf. A.6). (One may further extend these definitions to operators on sections of vector bundles, but there is no index formula in this context because of an ambiguity on the definition of $\mathcal{O}$.)

(3.7) Let X, Σ, $\mathcal{O}$ be as in definition 3.1, and let A be any Fourier integral operator on X whose canonical relation induces the identity map on Σ. Then the operator $T_A : \mathcal{O}^s \to \mathcal{O}^{s-m}$ ($m = \deg A$) is a Toeplitz operator. This follows from the microlocal analysis of (3.4): with the notations of (3.4), H^*T_AH is (microlocally) a pseudodifferential operator on R^N, so there exists a Toeplitz operator T_Q such that $H^*T_QH \sim H^*T_AH$ hence $T_Q \sim T_A$. Pasting together one gets globally a pseudodifferential operator Q such that $T_Q \sim T_A$, i.e. $T_Q - T_A$ sends $\mathcal{O}^s$ into $\mathcal{O}^\infty$ for any s. Now if R is any operator such that $R(\mathcal{O}^s) \subset \mathcal{O}^\infty$ for all s, R is a Toeplitz operator of degree $-\infty$ (we have $R = T_Q$ where Q is the operator on distributions of X defined by $Qu = R(Su)$; Q is smoothing, so it is a pseudodifferential operator of degree $-\infty$).

(3.8) Let X, Σ, $\mathcal{O}$, X', Σ', $\mathcal{O}'$ be as in definition 3.1, $\chi : \Sigma \to \Sigma'$ an isomorphism of symplectic cones and B a Fourier integral operator adapted to χ. Then $A = S'B$ is also adapted to χ and it follows from (3.7) that SA^*A induces an elliptic Toeplitz operator (of degree $2m$, $m = \deg A$). The operator $u \mapsto Au$ from $\mathcal{O}^s$ to $\mathcal{O}'^{s-m}$ is then elliptic; it has a parametrix, which is induced by a Fourier integral operator of degree $-m$ adapted to χ^{-1}, and has an index.

It follows that, up to an elliptic quasi-isomorphism, the Toeplitz space $\mathcal{O}$ and the Toeplitz algebra only depend on Σ and not on the embedding of Σ in T^*X.

(3.9) We now look more specially at the case where X is an oriented compact contact manifold, Σ the symplectic cone defining the contact structure. In this case we will also say that a Toeplitz structure associated with Σ is a *quantized contact structure* on X.

Let X, X' be two quantized contact manifolds, S and S' the corresponding projectors, and $\chi : X \to X'$ an oriented contact isomorphism; we will still denote by χ the extension $\Sigma \to \Sigma'$ to the symplectic cones defining the contact structures. Let P be a pseudodifferential operator on X, of degree m, and let $A: \mathcal{O}^s \to \mathcal{O}'^{s-m}$ be defined by $Au = S'(Pu_o\chi^{-1})$; it is induced by a Fourier integral operator adapted to χ; A is elliptic and admits a parametrix of degree $-m$ if P is elliptic on Σ. If B is a parametrix and T_Q Toeplitz operator on X, AT_QB is a Toeplitz operator on X′ with symbol $\sigma(T_Q)_o\chi^{-1}$. Operators such as A play in our context the same role as Fourier integral operators associated with real symplectic isomorphisms.

(3.10) Let X be the boundary of a strictly pseudo-convex bounded complex domain, Σ the symplectic cone defining its canonical oriented contact structure. Then the Szegö projector is a Fourier integral operator adapted to Id_Σ (cf. [4]), so X is canonically equipped with a quantized contact structure; the corresponding subspace is the space of boundary values of holomorphic functions. The Toeplitz operators defined in [2] are the Toeplitz operators in the sense above.

A.4. *Existence of Toeplitz structures*

Let X be a C^∞ compact manifold, $\Sigma \subset T^*X - 0$ a smooth symplectic subcone of the cotangent bundle of X (minus the zero section). In this section we prove the following result:

THEOREM 4.1. *There exists a Toeplitz structure associated with* Σ.

In particular if X is a compact oriented contact manifold, there exists a quantized contact structure on X. We will prove further:

THEOREM 4.2. *Let* X *be a compact oriented contact manifold,* G *a compact group of contact transformations of* X. *There exists a* G-*invariant quantized contact structure on* X.

We will prove these results in three steps. The first step is to construct globally the canonical relation associated with the projector S; it is based on Lemma 4.4 which we prove at the end of this section (cf. also [8]). First we need a definition.

DEFINITION 4.3. *Let* Y *be a smooth symplectic manifold,* $\Sigma \subset Y$ *a symplectic submanifold. We say that an ideal* $I \subset C^\infty(Y, \mathbb{C})$ *is* $\gg 0$ *relative to* Σ *if*

(i) I *is stable by Poisson bracket, and* Σ *is the set of common real zeros of functions* $f \in I$.

(ii) *Locally* I *is generated by* d *functions* $\zeta_1, \cdots, \zeta_d$ *(where* $d = \frac{1}{2}$ codim Σ) *such that the hermitian matrix* $\frac{1}{i}\{\zeta_p, \bar{\zeta}_q\}$ *is* $\gg 0$ *on* Σ.

LEMMA 4.4. *Let* Y *be a smooth symplectic manifold,* $\Sigma \subset Y$ *a symplectic submanifold, and* q *a real* C^∞ *function such that* $q = 0$ *on* Σ, $q > 0$ *outside of* Σ, *and* q *is transversally elliptic along* Σ *(i.e. the transversal hessian is non-degenerate). Then there exists a unique ideal* $I \gg 0$ *such that* $q \in I$.

EXAMPLE. Let X be the boundary of a strictly pseudo-convex complex domain, $\Sigma \subset T^*X - 0$ the symplectic half-line bundle attached to it (A.1), Y an open conic neighborhood of Σ in T^*X excluding $-\Sigma$, $q = q(x, \xi) = \Sigma|\langle \Xi_j(x), \xi \rangle|^2$ where the Ξ_j are antiholomorphic vector fields tangent to X such that any antiholomorphic vector field tangent to X is a linear combination of the Ξ_j (with C^∞ coefficients). Then I is the ideal generated by the functions $(x, \xi) \mapsto \langle \Xi_j(x), \xi \rangle$ (i.e. by the symbols of all antiholomorphic vector fields tangent to X).

LEMMA 4.5. *Let* Y, Σ, I *be as in Lemma 4.4, and let* V *be the almost complex manifold defined by* I. *There exists a unique almost complex* $\gg 0$ *canonical relation* $\mathcal{C}$ *in* $Y \times Y$ *such that*

$$\operatorname{diag} \Sigma \subset \mathcal{C} \subset V \times \bar{V} .$$

Proof. Since I is stable by Poisson brackets, V is involutive and its hamiltonian flow is well defined. $\mathcal{C}$ is then precisely the hamiltonian flow-out of $V \times \overline{V}$ from diag Σ (one checks immediately that this is a $>> 0$ canonical relation).

In the example above, $\mathcal{C}$ is the canonical relation corresponding to the Szegö projector (cf. [4]). Here is another example.

Let $Y = T^*R^{p+q} \supset \Sigma = T^*R^p$, where we denote the variables by (x, y, ξ, η); Σ is defined by $y = \eta = 0$. Let $q = \eta^2 + |\xi|^2 y^2$ (where we have set $y^2 = \Sigma y_j^2$). Then I is the ideal generated by the functions $\eta_j - i|\xi|y_j$, and $\mathcal{C}$ is the canonical relation corresponding to the phase function $<x-x', \xi> + \frac{1}{2} i|\xi|(y^2 + y'^2)$; its equations are $x - x' + \frac{1}{2} i(y^2 + y'^2) \xi/|\xi| = 0$, $\eta = i|\xi|y$, $\eta' = -i|\xi|y'$. It follows from [1], §10 that the situation of Lemma 4.5 is always locally isomorphic to that of the example above, with $\dim Y = 2p + 2q$, $\operatorname{codim} \Sigma = 2q$, which yields another proof of Lemma 4.5.

COROLLARY 4.6. *With the notations above we have* $\mathcal{C} \circ \mathcal{C} = \mathcal{C}^* = \mathcal{C}$.

Indeed this is true for the canonical relation of the Szegö projector, and also in the second example above, which is locally universal (it is also easy to verify it directly).

Let us now complete step one. We apply Lemma 4.4 and 4.5 with $Y = T^*X - 0$, Σ the given symplectic cone, and q homogeneous, ≥ 0 and transversally elliptic as in Lemma 4.4 (such a function always exists). Then because of the uniqueness, I and $\mathcal{C}$ are homogeneous. If there is a compact group G acting, we choose q G-invariant (this is possible, replacing q by its mean over G if need be). Then again I and $\mathcal{C}$ are G-invariant. We will construct S as a Fourier integral operator associated with $\mathcal{C}$.

Step 2. Let us denote by $\mathcal{L}^o_{\mathcal{C}}$ the set of Fourier integral operators of degree 0 associated with $\mathcal{C}$. If $A, B \in \mathcal{L}^o_{\mathcal{C}}$, it follows from Corollary 4.6 that $A_o B \in \mathcal{L}^o_{\mathcal{C}}$ and $A^* \in \mathcal{L}^o_{\mathcal{C}}$. Further since $\mathcal{C}$ induces the identity map on diag Σ, the restrictions to diag Σ of elements of $\mathcal{L}^o_{\mathcal{C}}$ are canonically identified with functions on diag Σ, and we have $\sigma(A_o B) =$

$\sigma(A)\sigma(B)$, $\sigma(A^*) = \overline{\sigma(A)}$ on diag Σ. We now choose A such that $\sigma(A) = 1$ on diag Σ, and $A = A^*$ (replacing it by $\frac{1}{2}(A+A^*)$ if need be). Since the symbol of A is elliptic on diag Σ, A is adapted to Id_Σ (in the sense of definition 2.7).

Step 3. Let A be as in step 2. Since the symbol of $A^2 - A$ vanishes on diag Σ, $A^2 - A$ is of degree $\leq -\frac{1}{2}$ as a Hermite operator, so it is a compact operator on $L^2(X)$. It follows that the spectrum of A consists of a sequence of real points, with 0 and 1 as only accumulation points. Let now F be an analytic function in a neighborhood of the spectrum of A such that F takes only the values 0 or 1 $(F^2 = F)$, F is real (i.e. $\overline{F}(\overline{z}) = F(z)$), and F is equal to 0 near $z = 0$ and to 1 near $z = 1$. Then clearly $S = F(A)$ is an orthogonal Fourier integral projector adapted to Id_Σ, and it defines a Toeplitz structure as required.

For Theorem 2, we may suppose A G-invariant (replacing it if need be by a mean over G, which will preserve the fact that $A = A^*$ and $\sigma(A) = 1$ on diag Σ). Then $S = F(A)$ is also G-invariant.

We end this section with a proof of Lemma 4.4 (which is also proved by Sjostrand and Menikoff in [8]). Let us first notice that the condition $I >> 0$ implies that I is completely determined by its Taylor expansion along Σ, i.e. a function vanishing of infinite order on Σ belongs to I. Also, because of the uniqueness, the result is local and we may restrict ourselves to arbitrarily small neighborhoods of points of Σ (the resulting ideals will paste back together).

We will prove by induction that I is uniquely determined mod. I_Σ^N for all N, I_Σ denoting the ideal of functions which vanish on Σ; this gives a construction of I by successive approximations and proves the uniqueness.

For $N = 2$ we use the following result, which is also used in [1] and [9]:

LEMMA 4.7. *Let* E *be a real symplectic vector space of dimension* 2d, σ *its symplectic form,* $\{\ \}$ *the dual form on the dual* E^*, Q *a* $>> 0$

quadratic form on E. *There exist complex linear forms* $z_1, \cdots, z_d$ *on* E, *unique up to linear combinations, such that* $\{z_p, z_q\} = 0$, $\frac{1}{i}\{z_p, \bar{z}_q\}$ *is hermitian* $>> 0$, $Q = \Sigma \lambda_{pq} z_p \bar{z}_q$ *for some positive hermitian matrix* (λ_{pq}).

The subspace $\Lambda \subset \mathbf{C} \otimes E^*$ spanned by the z_j is the unique $>> 0$ Lagrangian subspace which is isotropic for Q^{-1}. If we write the polar form of Q $Q(u, v) = \sigma(Au, v)$ with $A \in L(E)$, A is antisymmetric (both for σ and Q) so its eigenvalues go by pure imaginary opposite pairs. The orthogonal $\Lambda^{\perp} \subset \mathbf{C} \otimes E$ is spanned by the eigenvectors of A corresponding to eigenvalues with positive imaginary part. It depends smoothly on Q and σ so the construction still works for smooth vector bundles.

In our situation, the subspace spanned by the $df(x)$, $f \in I$, $x \in \Sigma$ is uniquely determined: it is the subspace Λ given by Lemma 4.7, with $E = T_x X / T_x \Sigma$, Q the quadratic form induced by the hessian of q. It follows that I is uniquely determined mod. I_Σ^2.

Suppose now that we have determined I mod. I_Σ^N. Then I is generated mod. I_Σ^N by d functions $\zeta_1, \cdots, \zeta_d$ such that the matrix $-i\{\zeta_p, \bar{\zeta}_q\}$ is hermitian $>> 0$ on Σ. Replacing the ζ_j by suitable linear combinations with C^∞ coefficients we may even suppose

$$-i\{\zeta_p, \bar{\zeta}_q\} = \delta_{pq} \text{ (the Kronecker symbol) on } \Sigma .$$

The condition on I means that we have

$(i)_N$ $\{\zeta_p, \zeta_q\}$ is a linear combination of the ζ_j mod. I_Σ^{N-1}.

In other words we have

$$\{\zeta_p, \zeta_q\} = \sum_{|\alpha|=N-1} \mu_{pq}^{\alpha} \bar{\zeta}^{\alpha} + \text{combination of the } \zeta_j .$$

$(ii)_N$ q is a linear combination of the ζ_j mod. I_Σ^{N+1}.

In other words we have

$$q = \Sigma v_j \zeta_j + \sum_{|\alpha|=N+1} q^{\alpha} \bar{\zeta}^{\alpha} .$$

Since q is real and its transversal hessian non-degenerate, we have $v_j = \Sigma\lambda_{jk}\bar{\zeta}_k$ mod. I_Σ^2 for some hermitian matrix $(\lambda_{jk}) >> 0$ on Σ.

We will now look for new functions $\zeta_j' = \zeta_j + \sum_{|a|=N} \omega_j^a \bar{\zeta}^a$ improving N into N+1. It will be convenient to introduce the following polynomial forms and vector fields in the formal variables $Z_1, \cdots, Z_d$, whose coefficients are smooth functions on Σ:

$$\mu = \Sigma\, \mu_{pq}^a Z^a dZ_p \wedge dZ_q$$

$$\omega = \Sigma\, \omega_j^a Z^a dZ_j$$

$$\phi = \Sigma\, q^a Z^a$$

$$V = \Sigma\, \lambda_{jk} Z_k \partial/\partial Z_j .$$

We have, with $r_j = \sum_a \omega_j^a \bar{\zeta}^a$

$$\{\zeta_p', \zeta_q'\} = \{\zeta_p, \zeta_q\} + \{\zeta_p, r_q\} - \{\zeta_q, r_p\} + \{r_p, r_q\} + \text{combination of the } \zeta_j .$$

Now $\zeta_j - \zeta_j' \in I_\Sigma^N$, so being a combination of the ζ_j or the ζ_j' mod. I_Σ^N is the same thing. We also have $\{r_p, r_q\} \in I^{2N-2} \subset I^N$ since $N \geq 2$, and since $\{\zeta_p, \bar{\zeta}_q\} = i\delta_{pq}$ on Σ, $\{\zeta_p, a\bar{\zeta}^a\} = i\, a \dfrac{\partial\bar{\zeta}^a}{\partial\bar{\zeta}_p}$ mod. I_Σ^N if $|a| = N$. Thus the condition $(i)_{N+1}$ for the ζ_j' is

$$\Sigma\, \mu_{pq}^a \bar{\zeta}^a + i\,\Sigma\, \omega_q^a \frac{\partial\bar{\zeta}^a}{\partial\bar{\zeta}_q} - i\,\Sigma\, \omega_p^a \frac{\partial\bar{\zeta}^a}{\partial\bar{\zeta}_q} \in I_\Sigma^N$$

or equivalently

(a) $\mu + i\, d\omega = 0$.

We next have $q = \Sigma v_j \zeta_j - \Sigma v_j r_j + \Sigma q^a \bar{\zeta}^a$, and since $v_j = \Sigma\lambda_{jk}\bar{\zeta}_k$ mod. I_Σ^2, the condition $(ii)_{N+1}$ for the ζ_j' is $\Sigma v_j r_j = \Sigma q^a \bar{\zeta}^a$ mod. I_Σ^{N+2} or equivalently

(b) $V \llcorner \omega = \phi$.

Now it follows from the Jacobi identity for Poisson brackets that we have $d\mu = 0$, so equation (a) has a solution. We are thus reduced to solve $d\omega = 0$, $V \llcorner \omega = \phi'$ (with ϕ' a homogeneous polynomial of degree N+1 of Z as ϕ). This is equivalent to $\omega = d\psi$, $V(\psi) = \Sigma\lambda_{jk} Z_k \partial\psi/\partial Z_j = \phi'$, with ψ a homogeneous polynomial of degree N+2 in Z. This finally has a unique solution because (λ_{jk}) is hermitian $>> 0$ (V acts on the space on homogeneous polynomials of degree N+2 as a differential operator, and its eigenvalues are the numbers $\alpha_1\lambda_1 + \cdots + \alpha_d\lambda_d$ with $|\alpha| = N+2$, $\lambda_1, \cdots, \lambda_d$ being the eigenvalues of (λ_{jk}): these are all > 0).

Thus I is uniquely determined mod. I^{N+1}, and by induction this ends the proof.

EXERCISE. If in Lemma 4.4 the data is analytic (i.e. Y a real analytic symplectic manifold, Σ an analytic symplectic submanifold, q a real analytic function), then the solution is analytic.

A.5. *Resolutions*

Let X be a smooth compact manifold, $\Sigma \subset T^*X - 0$ a symplectic subcone, $\mathcal{O}^o \subset L^2(X)$ defining a Toeplitz structure (definition 2.7), S the corresponding orthogonal projector. The purpose of this section is to describe the regular pseudodifferential equations satisfied by functions $f \in \mathcal{O}^o$, and the relations between these. When X is a contact manifold, Σ its symplectic cone, we will construct a complex $\overline{D}$ of first order pseudodifferential operators which will play for $\mathcal{O}^o$ the same role as $\bar\partial_b$ for boundary values of holomorphic functions.

We denote by $\mathfrak{D}$ the algebra of regular pseudodifferential operators on X, and $\mathfrak{J} \subset \mathfrak{D}$ the ideal of pseudodifferential operators P such that $P\mathcal{O}^o = 0$ (or equivalently $P_oS = 0$). $\mathfrak{D}$ is a filtered algebra, and the corresponding graded algebra is canonically isomorphic with the graded algebra $\mathfrak{A}$ of smooth homogeneous functions on $T^*X - 0$, via the symbol map. The symbol $I = \sigma(\mathfrak{J})$ is the graded ideal corresponding to $\mathfrak{J}$; it is generated by the symbols $\sigma(P)$, $P \in \mathfrak{J}$ in $\mathfrak{A}$, and it is stable by Poisson brackets.

PROPOSITION 5.1. *The ideal* $I = \sigma(\mathcal{J})$ *of symbols of pseudodifferential operators* P *such that* $PS = 0$ *is* $\gg 0$ *relative to* Σ *(definition 4.3).*

Let us first notice that if P is a regular pseudodifferential operator such that PS is of degree $-\infty$ $(PS \sim 0)$, then PS is a regular pseudodifferential operator and we have $P \sim P - PS$, $(P-PS)S = 0$. Thus I is also the symbol of the ideal of pseudodifferential operators P such that $PS \sim 0$, so we may microlocalize. We will denote by $\mathfrak{D}$ the sheaf (on T^*X-0) of pseudodifferential operators mod. operators of degree $-\infty$, operating on the sheaf of microfunctions, and $\mathfrak{J}$ the sheaf of ideals of operators P such that $PS \sim 0$. Now we have seen (3.4) that there exists microlocally near any point of Σ a local homogeneous isomorphism $\chi : T^*R^n \to \Sigma$ and a Fourier integral operator H adapted to χ such that $S \sim HH^*$. Let $\mathcal{C}'$ be the almost complex canonical relation of H (so the canonical relation of S is $\mathcal{C}' \circ \mathcal{C}'^*$), and $\mathcal{L}^m_{\mathcal{C}'}$ the set of Fourier integral operators of degree m belonging to $\mathcal{C}'$.

LEMMA 5.2. *Let* E, F *be two real symplectic vector spaces,* $\Lambda \subset \mathbf{C} \otimes E \times F$ *a Lagrangian subspace such that the first projection* $\mathrm{Re}\,\Lambda \to E$ *is bijective. Then the second projection* $\Lambda \to \mathbf{C} \otimes F$ *is injective, and the real part of its image is the second projection of* $\mathrm{Re}\,\Lambda$.

The hypothesis means that $\mathrm{Re}\,\Lambda$ is the graph of a real linear map $u : E \to F$. Since Λ is Lagrangian, u is symplectic up to a sign $(\sigma(ux, uy) = -\sigma(x, y))$, so it is injective and its range is a symplectic real subspace of F. Let E' be the range of u, G its orthogonal supplement so that F is identified with $E' \times G$. Then Λ is the set of vectors of the form (x, ux, y) for $x \in \mathbf{C} \otimes E$ and $y \in \Lambda'$, with Λ' a Lagrangian subspace of $\mathbf{C} \otimes G$. Since u is injective, the second projection $pr_2 : \Lambda \to \mathbf{C} \otimes F$ is injective; and since $\mathrm{Re}\,\Lambda$ is the graph of u, we have $\mathrm{Re}\,\Lambda' = 0$; in particular $pr_2\Lambda$ is transversal to its conjugate.

COROLLARY 5.3. *The second projection is an immersion of* $\mathcal{C}'$ *into* T^*X. *The image* $pr_2(\mathcal{C}')$ *is an almost complex submanifold of* T^*X, *transversal to its conjugate, whose set of real points is* Σ.

This follows from Lemma 5.2, with $E = -T(T^*R^n)$, $F = T(T^*X), \Lambda$ the tangent space of $\mathcal{C}'$.

COROLLARY 5.4. *Let* I' *be the ideal of smooth functions* f *on* T^*X *such that* $f \circ \mathcal{C}' = 0$ *(i.e.* f *vanishes on* $pr_2(\mathcal{C}')$*). Then* I' *is* $>> 0$ *relative to* Σ.

Proof. Let $\zeta_1 = \cdots = \zeta_d = 0$ be the equations of $pr_2(\mathcal{C}')$ $(d = \frac{1}{2}$ codim Σ $=$ codim $pr_2\, \mathcal{C}')$. Then the matrix $-i\{\zeta_p, \overline{\zeta}_q\}$ is nondegenerate because $pr_2(\mathcal{C}')$ meets its conjugate transversally along Σ and Σ is symplectic. It is positive because $\mathcal{C}'$ is $>> 0$.

COROLLARY 5.5. *Any smooth function* a *on* $\mathcal{C}'$ *is of the form* $p \circ pr_2$ *for some smooth function* p *on* T^*X.

This follows from the fact that $pr_2 : \mathcal{C}' \to T^*X$ is an immersion.

COROLLARY 5.6. *For any operator* $A \in \mathcal{L}^m_{\mathcal{C}'}$, *there exists a pseudodifferential operator* P *of degree* m *such that* $A \sim PH$.

Proof. Since H is elliptic of degree 0 there exists a smooth homogeneous function a of degree m on $\mathcal{C}'$ such that $\sigma(A) = a\,\sigma(H)$. It follows from Corollary 5.4 that there exists a pseudodifferential operator P of degree m on X such that $\sigma(A) = \sigma(PH)$, i.e. $A - PH$ is of degree $\leq m-1$ (as a Fourier integral operator). By successive approximations we may then improve so as to have $A \sim PH$.

We may now prove Proposition 5.1. Let first $P \in \mathfrak{J}$. Then $PH \sim 0$, so $\sigma(P) \in \mathbf{I}'$, and (locally) $\mathbf{I} \subset \mathbf{I}'$ ($\mathbf{I}$ and $\mathbf{I}'$ denote the sheafs generated by I and I', and for short we have written $\in$ instead of "is a section of"). Conversely let $q \in \mathbf{I}'$, and let Q be a pseudodifferential operator with symbol q, $m = \deg Q$ (Q is only microlocally defined). Then $\sigma_m(QH) = 0$ so QH is of degree $\leq m-1$ and there exists Q' of degree $\leq m-1$ with $QH \sim Q'H$. We then have $(Q-Q')H \sim 0$, i.e. $Q - Q' \in \mathfrak{J}$, and $\sigma(Q-Q') = \sigma(Q) = q$. It follows that we have $\mathbf{I} = \mathbf{I}'$, and Proposition 5.1 follows from Corollary 5.4.

From now on we suppose that X is an oriented contact manifold, Σ its symplectic half-line bundle, and that we are given a compact group G of contact transformations of X which leave invariant the given Toeplitz structure. Let $L \subset T^*X$ be the line bundle generated by Σ ($L = \Sigma \cup (-\Sigma) \cup$ the zero section of T^*X). T^*X/L is a real G-vector bundle on X whose pull back on Σ is canonically isomorphic with the normal tangent bundle of Σ $(T(T^*X)|\Sigma)/T\Sigma$.

We denote by V the G-complex vector bundle on X equal to T^*X/L equipped with the unique complex structure for which the differentials $d\zeta$, $\zeta \in I$, are $\mathbb{C}$-linear; this is well defined because Σ is a half-line bundle and I is generated by homogeneous functions; it is G-invariant because I is G-invariant. We let

$$a_o : T^*X \to V \tag{5.7}$$

be the canonical linear projection; it is G-equivariant.

PROPOSITION 5.8. *There exists a smooth equivariant map* $a : T^*X - O \to V$, *homogeneous of degree* 1, *such that*

(i) a *is tangent to* a_o *along* Σ, $a \neq 0$ *outside of* Σ, *and* $a(-\xi) = -a(\xi)$.

(ii) *the components of* a *generate* I *near* Σ.

(ii) means that, locally, if $e_1, \cdots, e_{n-1}$ is a basis of V (with $\dim X = 2n-1$, so $n-1 = \frac{1}{2}\operatorname{codim}\Sigma$ = fiber dimension of V), we have $a = \Sigma \zeta_j e_j$ with $\zeta_j \in I$. This being so the differentials $d\zeta_j$ are linearly independent at all points of Σ because a is tangent to a_o, so the ζ_j generate I locally (they generate I mod. I^2). Then globally any function $f \in I$ is of the form $b.a$ for some smooth section b from T^*X to the pull-back of the dual V'.

We now prove Proposition 5.8. It follows from the definition of a_o that its components belong to I mod. functions vanishing of order 2 on Σ. So we may choose a smooth fiber map $r : T^*X - O \to V$, homogeneous of degree 1, vanishing of order 2 on Σ, so that the components of $a_o + r$ belong to I (this is true locally hence globally by partition of the unity). We may also suppose $r(-\xi) = -r(\xi)$ and $|r| \leq \frac{1}{2}|a_o|$ for some G-invariant norm on V, since r vanishes of order 2 on Σ and its behavior away from Σ does not matter. Finally we may suppose r G-equivariant, replacing it if need be by its mean over G (this will not change the fact that the components of $a_o + r$ belong to I since I is G-invariant and a_o G-equivariant, nor the fact that r vanishes of order 2 on Σ, $r(-\xi) = -r(\xi)$, $|r| \leq \frac{1}{2}|a_o|$). Then $a = a_o + r$ has the required properties.

THEOREM 5.9. *There exists a* G-*equivariant complex* $\bar{D}$ *of regular symmetric pseudodifferential operators*:

$$0 \longrightarrow C^\infty(X) \xrightarrow{\bar{D}_o} C^\infty(X,V) \xrightarrow{\bar{D}_1} C^\infty(X,\Lambda^2 V) \cdots \xrightarrow{\bar{D}_{n-2}} C^\infty(X,\Lambda^{n-1} V) \longrightarrow 0$$

such that

(i) $\bar{D}_o S = 0$.

(ii) *The symbol* $\sigma(\bar{D})$ *is the exterior multiplication by* $a(\xi)$.

Complex means that we have $\bar{D}_j \bar{D}_{j-1} = 0$ $(j = 1, \cdots, n-1)$. Symmetric means that the $\bar{D}_j$ have the same symmetry as a differential operator: $\bar{D}_j \sim \tilde{\bar{D}}_j$ (cf. A.0).

It then follows from the analysis of [1], §8 and §9 that we have:

(5.10) $\bar{D}$ is subelliptic except in degree 0 and $n-1$. In particular its homology groups are of finite dimension, except in degree 0, $n-1$.

(5.11) $\mathcal{O}^o \subset H^o(\bar{D}) = \operatorname{Ker} \bar{D}_o$, and $\operatorname{Ker} \bar{D}_o / \mathcal{O}^o$ is finite dimensional.

(5.12) The range of $\bar{D}_{n-1}$ is closed; the orthogonal projector on its orthocomplement is a Fourier integral operator adapted to $\mathrm{Id}_{-\Sigma}$ (it is "carried" by $-\Sigma$); we will denote it by S_{n-1}.

(5.13) There exists a non-regular pseudodifferential operator E (or rather a sequence $E_j : C^\infty(X, \Lambda^j V) \to C^\infty(X, \Lambda^{j-1} V)$, $1 \le j \le n-1$) of degree $-\frac{1}{2}$ (in fact $E \in OPS_\Sigma^{-1,-1}$ with the notation of [1]) such that $E\bar{D} + \bar{D}E = Id - S - S_{n-1}$ –smooth projector of finite rank.

If $G = 1$, we may still improve $\bar{D}$ so that $\ker \bar{D}_o = \mathbb{C}$, and $H^j(\bar{D}) = 0$ for $1 \le j \le n-2$. However if G is infinite, this may be no longer possible.

The complex $\bar{D}$ is essentially the same thing as a projective resolution of the $\mathfrak{D}$-module $\mathfrak{D}/\mathfrak{I}$, hence the title of this section.

We will prove Theorem 5.9 in three steps. The first is to construct $\bar{D}_o$, the second to construct $\bar{D}$ mod. smoothing operators. Let $d = (d_o, \cdots, d_{n-2})$ be the symbol of $\bar{D} : d_j$ is the exterior multiplication by $a : \Lambda^j V \to \Lambda^{j+1} V$; it is equivariant.

Step 1. Let $\bar{D}'_o$ be any equivariant symmetric pseudodifferential operator from $C^\infty(X)$ to $C^\infty(X, V)$ such that $\sigma(\bar{D}'_o) = d_o$. Then $\bar{D}'_o S$ is of degree ≤ 0 as a Fourier integral operator, so there exists a pseudodifferential operator Q of degree ≤ 0 such that $\bar{D}'_o S \sim QS$ (this is true locally by Corollary 5.6 hence globally by a pseudodifferential partition of the unity). We may suppose Q symmetric since the microsupport of S is Σ so the behavior of Q away from Σ is irrelevant. Since $\bar{D}'_o S$ is equivariant, we may further suppose Q equivariant (replacing it if need be by its mean over G). We then set $\bar{D}_o = (\bar{D}'_o - Q)(Id - S) = \bar{D}'_o - Q - (\bar{D}'_o - Q)S$: this is a regular symmetric equivariant pseudodifferential operator, as $\bar{D}'_o$, Q, and $(\bar{D}'_o - Q)S \sim 0$; its symbol is d_o since Q and $(\bar{D}'_o - Q)S$ are of degree ≤ 0; and we have $\bar{D}_o S = 0$ since $(Id - S)S = 0$.

Step 2. We will construct recursively $\bar{D}'_j$, symmetric and equivariant, so that $\bar{D}'_o = \bar{D}_o$, $\sigma(\bar{D}'_j) = d_j$, $\bar{D}'_j \bar{D}'_{j-1} \sim 0$. We first construct $\bar{D}'_1$.

PROPOSITION 5.14. *Let* P *be any pseudodifferential operator of degree* m *such that* $PS \sim 0$. *There exists a pseudodifferential operator* Q *of degree* $m-1$ *such that* $P \sim \bar{D}_o Q$.

Proof. We have $\sigma(P) \epsilon I$ so it follows from Proposition 5.8 that there exists a symbol q of degree $m-1$ such that $\sigma(P) = q \cdot d_o$. Then if $\sigma(Q) = q$ we still have $(P-Q\bar{D}_o)S \sim 0$, and $P-Q\bar{D}_o$ is of degree $\leq m-1$. By successive approximations we may improve Q to get $P \sim Q\bar{D}_o$. This still works if P acts on sections of vector bundles. Since $\bar{D}_o$ is symmetric and G-equivariant we may choose Q symmetric and equivariant if P is so (replacing it if need be by the mean over G of $\frac{1}{2}(Q+\tilde{Q})$).

Let now $\bar{D}''_1$ be any symmetric equivariant pseudodifferential operator with symbol d_1. Then $\bar{D}''_1\bar{D}_oS = 0$ and since $\bar{D}''_1\bar{D}_o$ is of degree ≤ 1 (because $d_1 d_o = 0$) there exists Q symmetric and equivariant of degree ≤ 0 such that $\bar{D}''_1\bar{D}_o \sim Q\bar{D}_o$. We then set $\bar{D}'_1 = \bar{D}''_1 - Q$.

We now proceed to construct the other $\bar{D}'_j$.

LEMMA 5.15. *Let* P *be any smooth symbol such that* $p \cdot d_{j-1} = 0$. *Then if* $j \geq 1$ *there exists a symbol* q *such that* $p = q \cdot d_j$.

Proof. Locally if $e_1, \cdots, e_{n-1}$ is a basis of local sections of V we have $a = \Sigma \zeta_j e_j$ where the ζ_j generate I, and the $d\zeta_j$ are linearly independent along $\Sigma_U - \Sigma$. Thus the problem has a formal solution along $\Sigma_U - \Sigma$ (this is the syzygie theorem for the formal Taylor series of the ζ_j). In other words there exists q' such that $p - q' \cdot d_j$ vanishes of infinite order on $\Sigma_U - \Sigma$. Now since the $d\zeta_j$, $d\bar{\zeta}_j$ are still linearly independent along $\Sigma_U - \Sigma$, the problem always has a solution if p vanishes of infinite order on $\Sigma_U - \Sigma$; for instance we may choose $q = p d_j^* |a|^{-2}$ where d_j^* is the adjoint of d_j for some smooth hermitian norm on V (we then have $d_j^* d_j + d_{j-1} d_{j-1}^* = |a|^2 \mathrm{Id}$, and $|a|^{-2}p$ is smooth and vanishes of infinite order on $\Sigma_U - \Sigma$). Thus our problem has local solutions, so by partition of the unity it also has global solutions. If p is of degree m, q is of degree $m-1$. And since d_j is symmetric and equivariant we may choose q symmetric and equivariant if p is so (symmetric meaning here that we have $p(-\xi) = (-1)^{\deg p} p(\xi)$).

COROLLARY 5.16. *Suppose* $\bar{D}'_j\bar{D}'_{j-1} \sim 0$ *if* $j < k$. *Then if* $0 < j < k$ *and if* P *is a pseudodifferential operator of degree* m *such that* $P\bar{D}'_{j-1} \sim 0$, *there exists a pseudodifferential operator of degree* $m-1$, Q, *such that* $P \sim Q\bar{D}'_j$.

Indeed let $p = \sigma(P)$. We have $p \cdot d_{j-1} = 0$ so there exists a symbol q such that $p = q \cdot d_j$. Then if $\sigma(Q) = q$, $P - Q\bar{D}'_j$ is of degree $\leq m-1$ and we still have $(P-Q\bar{D}'_j)\bar{D}'_{j-1} \sim 0$. By successive approximations we may improve Q to get $P \sim Q\bar{D}'_j$. Since the $\bar{D}'_j$ are symmetric and equivariant we may choose Q symmetric and equivariant if P is so.

Let us now suppose that we have constructed the $\bar{D}'_j$ for $j < k$ $(k \geq 2)$, and let $\bar{D}''_k$ be any symmetric equivariant operator with symbol d_k. Then we have $(\bar{D}''_k\bar{D}'_{k-1})\bar{D}'_{k-2} \sim 0$ and since $\bar{D}''_k\bar{D}'_{k-1}$ is of degree ≤ 1 $(d_k d_{k-1} = 0)$ there exists a symmetric equivariant pseudodifferential operator Q of degree ≤ 0 such that $\bar{D}''_k\bar{D}'_{k-1} \sim Q\bar{D}'_{k-1}$. We then set $\bar{D}'_k = \bar{D}''_k - Q$.

Step 3. We finally remove the mod. C^∞ ambiguity in the relations above. Suppose that the $\bar{D}'_j$ satisfy for some k, $0 \leq k \leq n-2$:

$$(C_j) \quad \bar{D}'_j\bar{D}'_{j-1} = 0 \quad \text{if} \quad j \leq k .$$

If $k = n-2$ we are finished. Otherwise we set $\bar{D}'_j = \bar{D}_j$ if $j \leq k$, and proceed to construct $\bar{D}_{k+1}$ so that $\bar{D}_{k+1} \sim \bar{D}'_{k+1}$, $\bar{D}_{k+1}\bar{D}_k = 0$.

LEMMA 5.17. *Let* F *be the orthonormal projector on the range of* $\bar{D}_k$. *Then* F *is a non-regular pseudodifferential operator of degree* $\frac{1}{2}$; *it is* G*-equivariant, and we have* $\bar{D}'_{k+1}F \sim 0$.

Proof. Let $\Box = \bar{D}_k^*\bar{D}_k + \bar{D}_{k-1}\bar{D}_{k-1}^*$. It follows from [1], §8 and §9 that there exists a non-regular pseudodifferential operator E of degree -1 and type $\frac{1}{2}$ ($E \in OPS_\Sigma^{-2,-2}$ with the notation of [1]) such that

(i) $E\Box = \Box E = \mathrm{Id} - \Pi$, where Π is the orthogonal projector on $\operatorname{Ker} \Box$. Π is smooth of finite rank if $1 \leq k \leq n-2$, and $\pi - S$ is smooth of finite rank if $k = 0$.

(ii) $E = E^*$, and $\operatorname{Ker} E = \operatorname{Ker} \Box$.

E is uniquely defined by these conditions so it is equivariant. Let $F = \overline{D}_k E \overline{D}_k^*$. We will show that F is the orthogonal projector on the range of $\overline{D}_k$. Since it is a non-regular pseudodifferential operator of degree 0 and type $\frac{1}{2}$ ($F \in OPS_\Sigma^{0,0}$) this will prove the lemma.

Since $E = E^*$, we have clearly $F = F^*$. We have $\operatorname{Ker} \overline{D}_k \supset \ker \square$ so it follows that $\overline{D}_k \square E = \overline{D}_k$. Now it follows from (C_k) that commutes with $\overline{D}_k^* \overline{D}_k$, hence so does E; also $\overline{D}_k \square = \overline{D}_k \overline{D}_k^* \overline{D}_k$, so

$$F \overline{D}_k = \overline{D}_k E \overline{D}_k^* \overline{D}_k = \overline{D}_k \overline{D}_k^* \overline{D}_k E = \overline{D}_k \square E = \overline{D}_k .$$

It follows that $F^2 = F \overline{D}_k E \overline{D}_k^* = \overline{D}_k E \overline{D}_k^* = F$. The relations $F = F^* = F^2 F \overline{D}_k = \overline{D}_k$, $F = \overline{D}_k (E \overline{D}_k^*)$ exactly mean that F is the orthogonal projector on the range of $\overline{D}_k$. They imply $\overline{D}'_{k+1} F = \overline{D}'_{k+1} \overline{D}_k E \overline{D}_k^* \sim 0$.

We now set $\overline{D}_{k+1} = \overline{D}'_{k+1}(\mathrm{Id} - F)$. We have $\overline{D}'_{k+1} F \sim 0$ so $\overline{D}_{k+1} \sim \overline{D}_{k+1}$. $\overline{D}_{k+1}$ is symmetric since $\overline{D}'_{k+1}$ is so, and we have $(\mathrm{Id} - F) \overline{D}_k = 0$ so $\overline{D}_{k+1} \overline{D}_k = 0$. This ends the proof.

A.6. *Final remarks*

(6.1) The definition above (2.7) of Toeplitz structures coincides with that of §2. Indeed let S be the projector defining a Toeplitz structure in the sense of 2.7, and $\mathcal{I}$ the ideal of pseudodifferential operators A such that $AS = 0$. Then the symbol ideal $I = \sigma(\mathcal{I})$ is $\gg 0$ relative to Σ so by [1], §10, there exists microlocally an elliptic Fourier integral operator $\mathcal{F}$ transforming $\mathcal{I}$ into the ideal generated by the $\partial/\partial y_j + y_j |D_x|$ $(j = 1, \cdots, d)$ on $\mathbf{R}^{p+d}$. In fact the operator $\mathcal{F}$ constructed in [1] is unitary ([1], Proposition 10.8), so if S_0 denotes the orthogonal projector on $\bigcap \operatorname{Ker}(\partial/\partial y_j + y_j |D_x|)$, we have $S \sim \mathcal{F}^* S_0 \mathcal{F}$ which is precisely the definition of §2.

(6.2) In the constructions above we have used Fourier integral operators with complex phases and regular symbols. This was to get the strongest possible statements about the Toeplitz algebra and in A.5. But for many constructions, e.g. for the index formula, or for the estimation of eigenvalues of elliptic operators, one may use weaker assumptions. For instance one may replace S by any Hermite projector whose symbol is a projector of rank one on a "vacuum state." Again for many purposes it is enough that the operator $\bar{D}$ of A.5 differs from a regular pseudodifferential operator by a non-regular operator of degree < 1. For instance we may simplify the construction of A.5 as follows: let first $\bar{D}'$ be any (regular) pseudodifferential operator whose symbol is the exterior multiplication by $a_0(\xi)$, where a_0 is the linear map: $T^*X \to V$ (5.8). Then $\bar{D}'_0 S$ is in fact of degree ≤ 0 as a Hermite operator; if we set $\bar{D}_0 = \bar{D}'_0(\mathrm{Id}-S)$, we have $\bar{D}_0 S = 0$, and again the range of S is of finite codimension in Ker $\bar{D}_0$. We may then perform directly the second step of the proof of Theorem 5.9 and get another complex $\bar{D}$ whose non-regular part is of degree $\leq \frac{1}{2}$, and whose symbol is the exterior multiplication by a_0.

(6.3) As was mentioned in A.2, an elliptic Toeplitz matrix has an index. This of course does not depend on the manner in which Σ is embedded in T^*X, or on the choice of the Toeplitz structure, but only on its symbol (because there exist adapted Fourier integral operators, which may be used to transport elliptic matrices, preserving the index and the symbol). We may thus suppose that X is a compact oriented contact manifold, Σ its symplectic half-line bundle. The index formula is the formula of [2], Theorem 1, and is proved as this theorem, using the complex $\bar{D}$ instead of $\bar{\partial}_b$.

(6.4) Let X be a smooth compact manifold, Y its cotangent sphere. Y is the basis of the symplectic cone T^*X-0 and is a smooth compact contact manifold whose symplectic half-line bundle $\Sigma \subset T^*Y-0$ is canonically isomorphic with T^*X-0. Let χ be the canonical isomorphism $T^*X-0 \to \Sigma$, A a Fourier integral operator adapted to χ. Then if S is any Fourier integral projector on Y adapted to Id_Σ, $\mathcal{O}^o$ its range ($\mathcal{O}^o = S(L^2(Y))$), the operator $u \mapsto SAu$ defines an elliptic transformation (of degree $m = \deg A$) from $H^s(X)$ to $\mathcal{O}^{s-m}$. It follows that the Toeplitz algebra on Y is essentially isomorphic with the algebra of pseudodifferential operators on X. A particular example of this is described in [3]: we may suppose that X is a real analytic riemannian manifold. Let $\tilde{X}$ be a complexification of X and $\phi = \phi(t,x,\xi) = (X(t,x,\xi), \Xi(t,x,\xi))$ be the geogesic flow (i.e. the hamiltonian flow of the function $|\xi|$). Since ϕ is analytic it is still well defined for small complex t. Let $\Omega_\varepsilon \subset \tilde{X}$ be the set of end points $X(it,x,\xi)$ with $0 \le t \le \varepsilon$. If ε is small enough, Ω_ε is strictly pseudo-convex and the map $(x,\xi) \mapsto X(i\varepsilon, x, \xi)$ defines a contact isomorphism χ from the cotangent sphere of X to $\partial\Omega_\varepsilon$. Finally if Δ is the Laplace operator of X, and $A = \exp(-\varepsilon|\Delta|^{1/2})$, A is a Fourier integral operator adapted to χ; in this case it is of degree $-\frac{n-1}{4}$, and it is a bijection from $L^2(X)$ to $\mathcal{O}^{-\frac{n-1}{4}}(Y)$, $\mathcal{O}$ denoting the space of boundary values of holomorphic functions.

PROPOSITION 6.6. *Let* X *be a compact manifold,* G *a compact group of homogeneous symplectic transformations of* T^*X-0. *Then* G *can be lifted as a compact group of Fourier integral operators on* X.

This result was proved by A. Weinstein. We prove it here as an application of (6.4). Let Y be the basis of T^*X-0; then G is a group of contact transformations of Y, and there exists a G-invariant Toeplitz structure on Y, associated with Σ, Σ being the symplectic half-line bundle of Y. Let us choose a Fourier integral operator A adapted to χ, χ being the canonical isomorphism $T^*X-0 \to \Sigma$: the range of S_0A is closed, of finite codimension in $\mathcal{O}^0(Y)$, so since G is compact there exists a G-invariant finite dimensional subspace $E \subset \mathcal{O}^\infty(Y)$ containing a supplement of the range of SA. Let us now set $B = (\mathrm{Id}-\Pi_E)SA$ (Π_E being the orthogonal projector on E). B is also adapted to χ so its kernel is finite dimensional, contained in $\mathcal{C}^\infty(X)$, and its range is closed, of finite codimension, and G-invariant, in $\mathcal{O}^0(Y)$. We may now define the action of G as the identity on $\mathrm{Ker}\, B$, and the pull-back by B of its action on the range of B, on $(\mathrm{Ker}\, B)^\perp$.

BIBLIOGRAPHY

[1] L. Boutet de Monvel, Hypoelliptic operators with double characteristics and related pseudodifferential operators. Comm. Pure Appl. Math., 27 (1974), 585-639.

[2] ———, On the index of Toeplitz operators of several complex variables. Inventiones Math., 50 (1979), 249-272.

[3] ———, Convergence dans le domaine complexe des séries de fonctions propres. C.R. Acad. Sc. Paris 287 (1978), 855-856.

[4] L. Boutet de Monvel and J. Sjöstrand, Sur la singularite des noyaux de Bergmann et de Szegö. Astérisque 34-35 (1976), 123-164.

[5] V. Guillemin, Symplectic spinors and partial differential equations. Coll. Inst. CNRS no. 237 Géométrie symplectique et physique mathématique, 217-252.

[6] G. Lebeau, Sur les systèmes holonomes à caractéristiques complexes. Thèse de troisième cycle, Orsay, 1978.

[7] A. Melin and J. Sjöstrand, Fourier integral operators with complex valued phase functions. Lecture Notes, Springer Verlag, no. 459, 120-223.

[8] A. Menikoff and J. Sjöstrand, On the eigenvalues of a class of hypoelliptic operators. Math. Ann. 235 (1978), 55-85.

[9] J. Sjöstrand, Parametrices for pseudodifferential operators with multiple characteristics. Arkiv för Mat. 12 (1974), 85-130.

Library of Congress Cataloging in Publication Data

Boutet de Monvel, L., 1941-
The spectral theory of Toeplitz operators.

(Annals of mathematics studies ; 99)
Bibliography: p.
1. Toeplitz operators. 2. Spectral theory (Mathematics)
I. Guillemin, V., 1937- . II. Title. III. Series.
QA329.2.B68 1981 515.7'246 80-8538
ISBN 0-691-08284-7 AACR2
ISBN 0-691-08279-0 (pbk.)

www.ingramcontent.com/pod-product-compliance
Ingram Content Group UK Ltd.
Pitfield, Milton Keynes, MK11 3LW, UK
UKHW031026161224
452563UK00001BA/102